PROTOPLASMATOLOGIA
HANDBUCH DER PROTOPLASMAFORSCHUNG

BEGRÜNDET VON
L. V. HEILBRUNN · F. WEBER
PHILADELPHIA   GRAZ

HERAUSGEGEBEN VON
M. ALFERT · H. BAUER · C. V. HARDING · W. SANDRITTER · P. SITTE
BERKELEY   TÜBINGEN   ROCHESTER   GIESSEN   HEIDELBERG

MITHERAUSGEBER

W. H. ARISZ-GRONINGEN · J. BRACHET-BRUXELLES · H. G. CALLAN-ST. ANDREWS
R. COLLANDER-HELSINKI · K. DAN-TOKYO · E. FAURÉ-FREMIET-PARIS
A. FREY-WYSSLING-ZÜRICH · L. GEITLER-WIEN · K. HÖFLER-WIEN
M. H. JACOBS-PHILADELPHIA · N. KAMIYA-OSAKA · W. MENKE-KÖLN
A. MONROY-PALERMO · A. PISCHINGER-WIEN · J. RUNNSTRÖM-STOCKHOLM

BAND VI
KERN- UND ZELLTEILUNG

G

DER KERNTEILUNGSMECHANISMUS

1

THE BEHAVIOR OF CENTRIOLES AND THE STRUCTURE AND FORMATION OF THE ACHROMATIC FIGURE

1966

SPRINGER-VERLAG WIEN GMBH

THE BEHAVIOR OF CENTRIOLES
AND THE STRUCTURE AND FORMATION
OF THE ACHROMATIC FIGURE

BY

HANS A. WENT
PULLMAN

WITH 30 FIGURES

1966

SPRINGER-VERLAG WIEN GMBH

ISBN 978-3-211-80783-5   ISBN 978-3-7091-5571-4 (eBook)
DOI 10.1007/978-3-7091-5571-4

TITEL-NR. 8747

Protoplasmatologia
  VI. Kern- und Zellteilung
    G. Der Kernteilungsmechanismus
      1. The Behavior of Centrioles and the Structure and Formation
      of the Achromatic Figure

The Behavior of Centrioles and the Structure and Formation of the Achromatic Figure

By

HANS A. WENT

Department of Zoology, Washington State University, Pullman, Washington, USA

With 30 Figures

Table of Contents

Introduction

Included among the many preparations for cell division in most higher plants and animals is the constellation of events encompassing the formation of the spindle or mitotic apparatus, depending upon the organism. These are transient structures, responsible primarily for the equipartition of the hereditary material, which are elaborated by cells just in advance of division and disappear from view shortly after (and sometimes before) the completion of division. Precisely what the origin and fate of the structural molecules of the division figure may be is apparently not known. In contrast to the dearth of factual information in this regard, an extensive literature describing the microscopically visible aspects of mitotic apparatus build-up and break-down exists. Recently investigations of a more chemical nature inquiring into the chemical events associated with mitotic apparatus appearance and disappearance have been reported. Whatever the underlying scheme may be, it seems likely that the formation of the division figure involves complex interactions between structural moieties that can be detected by light and electron microscopy, their structural precursors and functional molecules.

At the zenith of its structural development (usually metaphase) the mitotic apparatus is an impressively complex and highly ordered structure. Usually it is in a state of continous change from the moment it first becomes observable until it can no longer be detected, although in some cells development may be arrested at metaphase for weeks or months. During this period there appears to be no structural change and likely the turnover of the molecular constituents is not high. It is clear that the organization and activity of this dynamic structural complex is guided by some entity which itself must be under other cellular control mechanisms. The identity of this entity is not definitely known in most cases, but the centriole obviously fulfills this role in some cells, while in others spindle organization is clearly directed by the kinetochores.

MAZIA (1961: p. 81) views the complete picture of cell division as the sum of the p e r i o d o f d i v i s i o n and the p e r i o d b e t w e e n d i v i s i o n s. During the latter period occur most of the important processes that prepare the cell for the act of division. Although a sharp and useful demarcation can be constructed between these two periods, its physiological significance can be seriously questioned. The cell may not "recognize" these anthropomorphic designations in that division may be attempted when the cell is clearly unprepared for and incapable of division as the result of some exogenous trauma. Thus the inter-position of an insurmountable obstacle that renders it impossible for the cell to divide may not prevent the cell from attempting to do so. These obstacles, for example, may be an insufficient number of division centers or an improperly constructed mitotic apparatus.

This paper is not a review of the extant literature, which is voluminous, but only a consideration of selected, and it is hoped, the more pertinent literature directly applicable to the subject. The reader will be referred to more detailed accounts and reviews when they apply.

I. Behavior of Centrioles

A. Terminology and general structure

There appears to be considerable confusion and inconsistency associated with the term c e n t r a l   b o d y and/or its component units. WILSON (1928; p. 1126)states that the term "central bodies" is:

"a vague term designating the structures at the center of the aster, during mitosis. They include the minute c e n t r i o l e at the focus of the aster, and the larger c e n t r o s o m e by which it is surrounded. It is often difficult to determine whether a central body represents one or the other or both of these structures."

As indicated by WILSON (1928: p. 673) the central body was discovered by VAN BENEDEN (1876) who initially gave it the term p o l a r   c o r p u s c l e and later c e n t r a l   c o r p u s c l e (VAN BENEDEN and NEYT, 1887). BOVERI (1895) introduced the term c e n t r i o l e to designate the much smaller body within the centrosome. WILSON (1928: p. 673) illustrates various forms taken by a central body and its surrounding aster. The simplest configuration assumed by the central body is in the form of a single granule at the focus of the aster. In such cases WILSON feels justified to call the central body a centriole, although it has often been erroneously called a centrosome (WILSON, 1928: p. 675). In more complex situations he pictures the central body as consisting of the centriole surrounded by a sphere, the centrosome. The centrosome may be a more or less finely granular area displaying no preferential orientation of the granules or a radial sphere in which the radially oriented structures appear as centripital extensions of the astral rays. He, thus, clearly differentiates between centriole and centrosome, a distinction which is not always made in both the earlier and later literature.

In a footnote (p. 675) WILSON mentions that the word centrosome is used in at least four different senses in the literature.

GEITLER (1934: p. 47) schematically depicts the generalized scheme of the mechanical events of cell division in plants and animals. In this illustration, small dots which behave in a manner that would be expected of centrioles are designated centrosomes. However, in a footnote on p. 157 referring to this illustration, it is mentioned that the previously divided c e n t r i o l e s endure the succeeding interphase in anticipation of the following division. So both the terms centriole and centrosome have been used to designate the s a m e structure in the illustration on p. 47. A similar apparent confusion in terminology is to be found in COE (1899) in his fine description of maturation and fertilization in *Cerebratulus*. He uses centrosome to refer to the intensely staining spot or granule stained with Bordeaux red followed by iron-hematoxylin at the center of the aster. It would appear that this structure corresponds both to what would today be called a centriole and to Wilson's concept of the centriole.

The use of the above mentioned cases to illustrate the ill-defined usage of centriole and centrosome is intended only to make the reader cognizant of the fact that considerable confusion and ambiguity shrouds the use of these terms and is not intended to cast aspersion upon the observations by

1*

these (and other) investigators. It is clear that these words have been used synonymously and without adequate identification with the structure they are to designate.

Owing to the variability in usage, it should be apparent that when one encounters the terms centriole, centrosome, central body, etc. in the literature it should be carefully ascertained precisely what is the structure that is being referred to. Such caution will aid in reducing the continued use of these terms in an ambiguous manner. From the following discussion one can readily imagine how the confusion in terminology arose. BOVERI (1900) describes and illustrates complete sequences of changes undergone by the centrosome of *Ascaris megalocephala bivalens* during first cleavage. In one sequence the centrosome (Boveri's designation) enlarges from a single point (which approximately corresponds in size to a centriole) at metaphase to a large sphere at early anaphase. During the course of anaphase the enlarged centrosome becomes elongated, but with the onset of telophase it once again becomes spherical in shape and smaller in size. The size continues to decrease until sometime during interphase when the centrosome divides. As the sister centrosomes migrate to opposite poles of the nucleus, they again enlarge until a maximum size is attained at some stage of prophase. The centrosome retains its large size until metaphase at which time it begins to shrink and finally becomes concentrated to a mere point. BOVERI never refers to this structure as a centriole. In another illustrated sequence showing karyokinetic changes in similarly prepared material from the same animal, he detects the presence of two small granules which he calls centrioles in the elongated anaphase centrosome. From this it is quite clear that BOVERI could not consistently visualize centrioles (the differentiated body situated at the center of the aster) in his material and that he reserved the use of this term to those cases in which there could be no doubt of their identity. He suggests that in all mitotic stages in the center of the aster (Sphäre) there exist two concentrically oriented structures, the centrosome and the centriole, and that the centriole is a permanent body, persisting from division to division.

There exist many beautiful descriptions made prior to 1925 of the behavior of centrosomes (some of these are clearly centrioles) and centrioles in a large variety of cell types based upon classical cytochemical methods which will not be presented here. References to the early literature can be found in WILSON (1928), SHARP (1921), HERTWIG (1893, 1906) FISCHER (1899) and others. An excellent discussion and review of the behavior and morphology of centrioles, centrosomes, central bodies and basal granules can be found in HEIDENHAIN (1907).

The size and structure of centrioles as "seen" by light microscopy is variable. Excluding for the moment the Protozoa, the most frequently encountered form is that of an intensely staining sphere which in size is close to the limit of resolution of the microscope (e. g. BOVERI, 1900; CONKLIN, 1902; GEITLER, 1934; HEIDENHAIN, 1907; MEAD, 1898; POLLISTER, 1933; WILSON, 1897, 1898). Large centrioles in the form of curved rods (5 μ long) have been observed and described by COSTELLO (1961 a) for *Polychoerus*. They also can assume the straight rod form seen by JOHNSON (1931) in the cricket *Oecanthus nigricornis,* by SCHRADER (1941) in the earwig *Anisolabis martima,* by SCHREINER and SCHREINER (1905) in the hagfish *Myxine glutinosa* (L), and by MINOUCHI (1936) in the trematode *Polystomum.* Centrioles have also been described as being V, L and T shaped. These forms very likely do not represent the basic shape of the centriole, but are the result of the

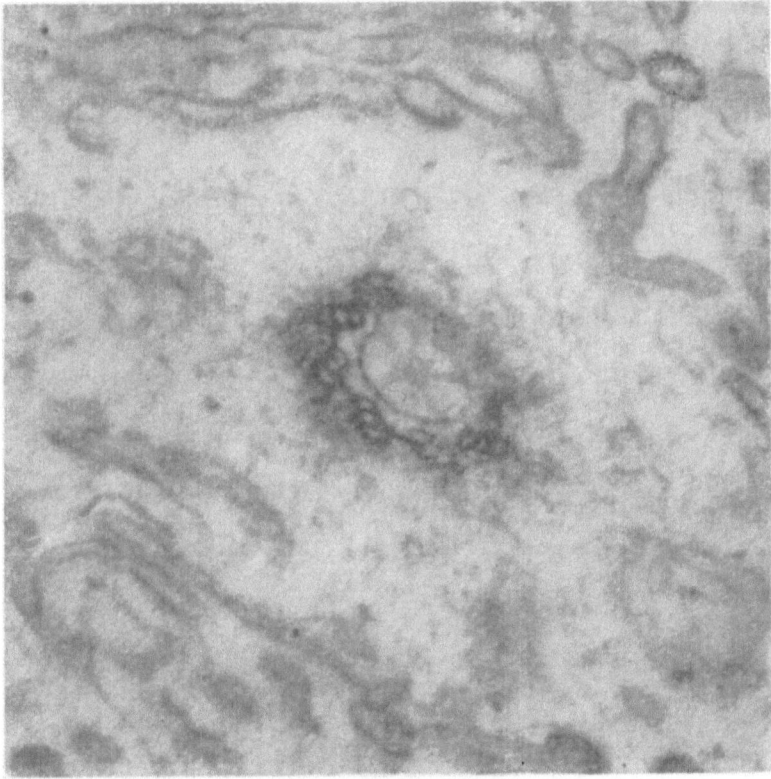

Fig. 1. Transverse section through a mature centriole from a very early atypical spermatocyte in *Viviparus.* The triplet fibers of the centriole wall are clearly visible. This photograph shows the central "hub" and the radial "spokes" imparting the cartwheel appearance to the core of the centriole. X 140,000. From GALL, J. Biophys. Biochem. Cytol. 10, 1961, 163.

configuration assumed by the rod shaped daughter centriole and the rod shaped parent centriole under whose influence the former is being generated. Certain flagellates (e. g. *Barbulanympha, Macrospironympha, Trichonympha, Pseudotrichonympha, Joenia, Gigantomonas,* etc.) possess very large and morphologically diverse centrioles. The variation in centriole structure among these flagellates (CLEVELAND, 1957 a, 1960 a, 1961, 1962) probably exceeds that for all other organisms combined. The morphology not only varies with the species, but can undergo drastic changes during the life cycle of the flagellate (CLEVELAND, 1957 a). In most organisms the centriole is at the center of the centrosome whenever both can be identified. However, in the flagellates the relationship may be modified. For example, in *Barbulanympha* the old centriole is a long rod shaped structure, whose length exceeds the diameter of the centrosome by several fold. The centrosome, from which all astral rays and spindle fibers emanate, is located at the posterior end of the old centriole. The anterior end of the old centriole is connected by a fine fiber to the new centriole, which begins its existence as a mere granule. Thus, one end of the centriole is the focus for the centrosome and organizes the fibrillar structure at one pole of the achromatic figure. The other, or generative, end is concerned with the production of

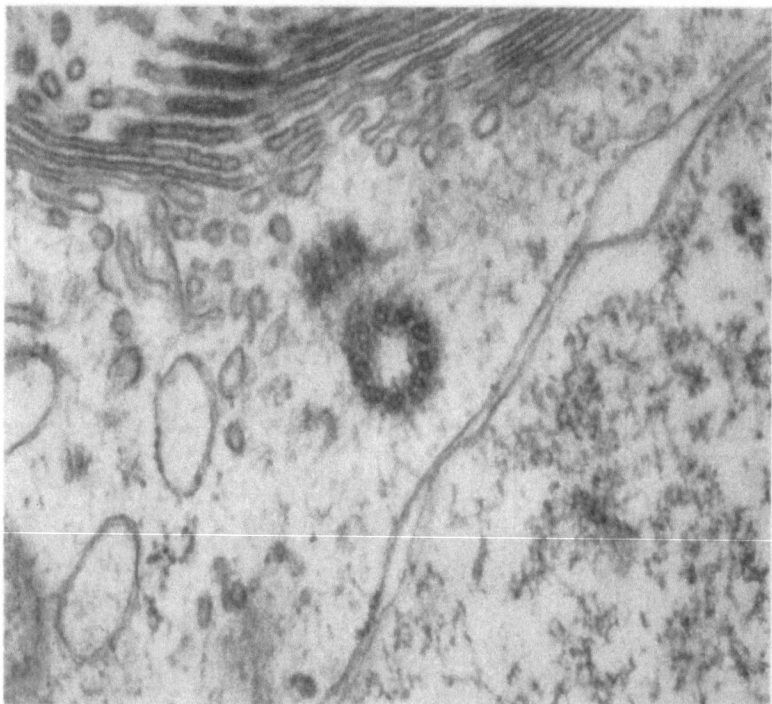

Fig. 2. Transverse section through a mature centriole and a longitudinal section through the edge of its accompanying procentriole in a primary spermatocyte (pachytene stage) of the typical series in *Viviparus*. The nine triplet fibers of the mature centriole can clearly be seen and there is the suggestion of fibers or tubules in the procentriole. X 75,000. From GALL, J. Biophys. Biochem. Cytol. 10, 1961, 163.

the new centriole and extra-nuclear organelles (CLEVELAND, 1961, 1962). In the centriole life cycle observed in *Macrospironympha*, the centrosome formed at the posterior end of the centriole becomes completely dissociated from the centriole and moves posteriorly, becoming associated with the nucleus as it migrates.

Since it is currently possible, with the advent of ultrathin sectioning techniques (PORTER and BLUM, 1953), to resolve the fine structure of centrioles, they can frequently be identified with greater certainty than had hitherto been possible. See Figures 1 through 4. When referring to other than electromicrographs, the term centriole, as used in this article, will identify those structures which conform to the concept proposed by BOVERI (1900) and as employed by WILSON (1928). The electron microscope, however, allows one also to make a precise positive identification of centrioles based upon their highly constant fine structure. SLEIGH (1962) presents a brief, but good discussion of centrioles and basal bodies. See also CARASSO and FAVARD (1961). The literature contains a number of electronmicrographs

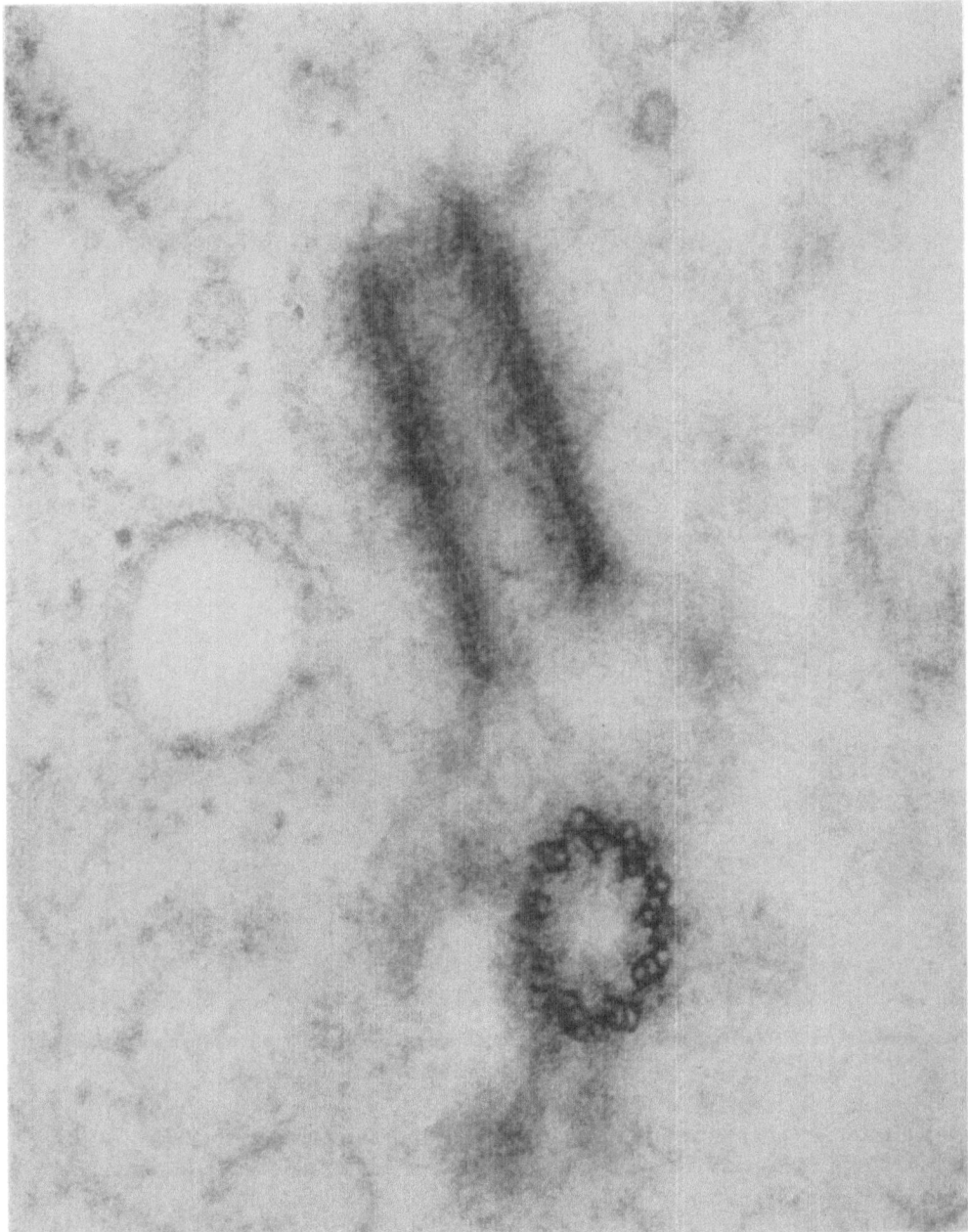

Fig. 3. Two mature centrioles, one cut transversely and the other longitudinally in a human Lymphosarcoma cell. The two centrioles are still approximately perpendicular to each other. If these are parent and daughter centrioles, the one cut longitudinally would probably represent the daughter centriole. From the transversely cut centriole there radiates two or three structures that may correspond to the pericentriolar bodies described by BESSIS and BRETON-GORIUS (1958). Courtesy of W. BERNHARD.

showing the fine structure of the centriole (BESSIS and BRETON-GORIUS, 1957 a:
BERNHARD and DE HARVEN, 1960; GALL, 1961; NAGANO, 1961). GALL (1961) offers
a pertinent discussion of centriolar structure accompanied by excellent

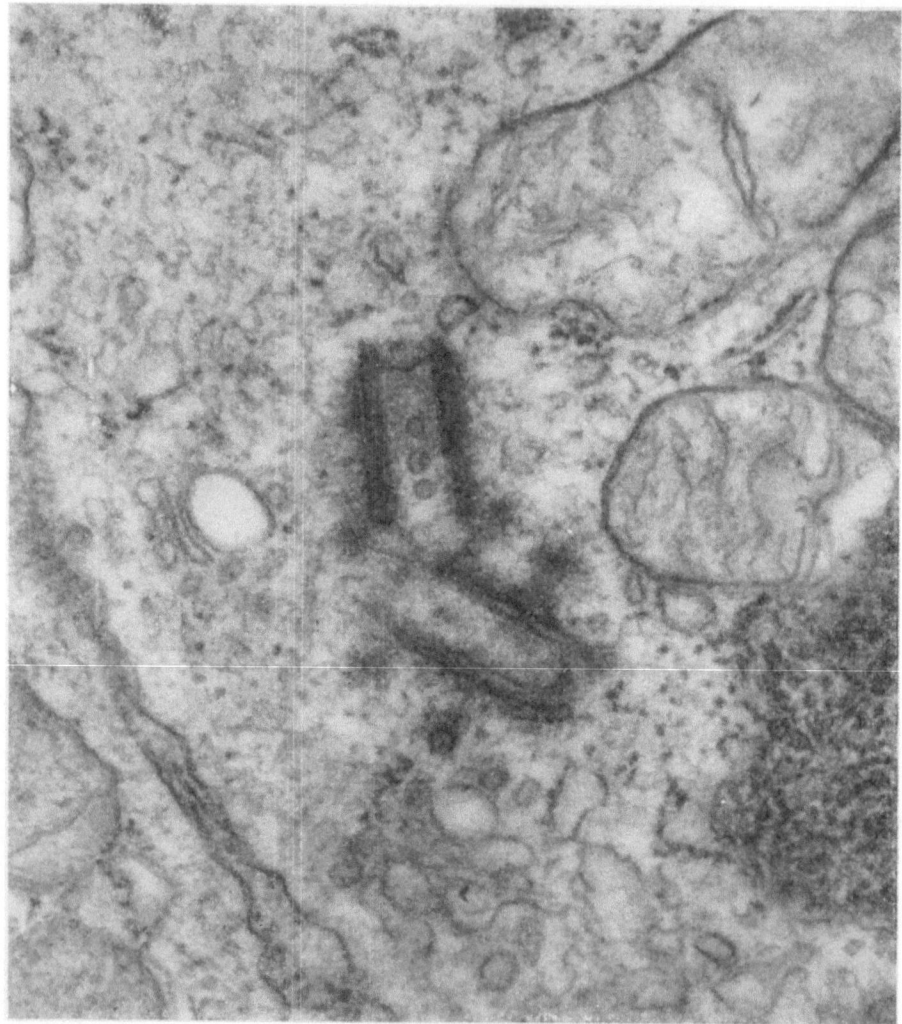

Fig. 4. Two mature centrioles in human lymphoid leukemia, both cut longitudinally, which are no longer perpendicular to each other. There appear to be pericentriolar bodies, or structures resembling them, associated
with one of the centrioles. Courtesy of W. BERNHARD.

electronmicrographs of the fine structure; and the criteria set forth by him
will serve as the reference point of this paper for centriolar structure. There
are electronmicrographs of structures identified as centrioles in which not
all of their fine structure (if, indeed they possess any) has been resolved
(SOTELO and TRUJILLO-CENOZ, 1958; NAGANO, 1959; HARRIS, 1962 a, 1962 b)
yet their general appearance and position leave little room for doubting
that they are indeed centrioles. In the most favorable electronmicrographs

(Bessis, Breton-Gorius and Thiery, 1958; Bessis and Breton-Gorius, 1957 b; Bernhard and de Harven, 1960; Gall, 1961) centrioles are hollow cylindrical assemblies of nine, usually three-tubed, elements oriented parallel to the long axis of the centriole (Figure 3). These three-tubed elements are referred to as triplet-fibers (Gall, 1961; Gibbons and Grimstone, 1960 — in basal bodies). In general, the centrioles are 300–500 mμ long, 120–160 mμ in diameter (Amano, 1957; Fawcett, 1961; Gall, 1961) and open at both ends. Sleigh (1962) presents a table giving the dimensions of centrioles from various sources. Most of the centrioles were 120–250 mμ in diameter by 200–750 mμ long.

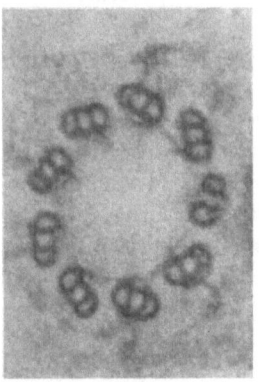 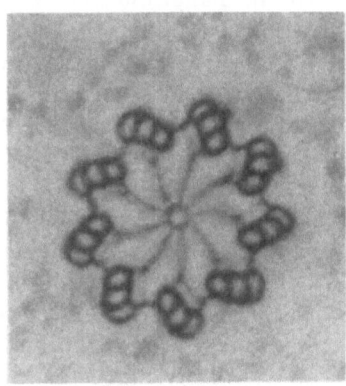

Fig. 5. Fig. 6.

Fig. 5. Transverse section through the distal region of a basal body of *Trichonympha*. At this level the lumen appears structureless. Observe the resemblance to the transversely cut centriole in Figure 3. X 150,000. From Gibbons and Grimstone, J. Biophys. Biochem. Cytol. 7, 697.

Fig. 6. Transverse section through the proximal region of a basal body of *Trichonympha*. At this plane the cartwheel structure in the lumen can be clearly seen. The axial planes of the triplet fibers are inclined in the same direction and at approximately the same angle. This is the orientation (clockwise) observed when looking along the basal body from the proximal to the distal end. The innermost subfiber of each triplet is subfiber A. Only clockwise oriented basal bodies have been seen in *Trichonympha*. X 150,000. From Gibbons and Grimstone, J. Biophys. Biochem. Cytol. 7, 697.

In the better electronmicrographs the core can be seen to contain delicate, orderly arranged, electron-dense material that imparts a cartwheel appearance to cross-sections of centrioles (Gall, 1961 — in *Viviparus* sperm; Gibbons and Grimstone, 1960 — basal bodies of *Paramecium* and *Trichonympha*). See Figures 1 and 6. The presence or absence of the cart-wheel appearance can also reflect the level at which the centriole or basal body was cut. See Figures 5 and 6. Frequently, however, the core appears devoid of any organized structure, although the tubular subfibers of the triplet-fibers may be resolved. In *Viviparus (Gall*, 1961) each triplet fiber contains three tubular-appearing subfibers. The less electron-dense core of each subfiber is about 12–13 mμ in diameter and bounded by a wall about 5 mμ thick. The overall dimensions, in cross-section, of each triplet fiber are about 23 mμ by 47 mμ. The triplet fibers are inclined at an angle of about 40⁰ from a tangent to the circumference of the centriole.

At the base of every cilium and flagellum is a small granule or short rod (Fawcett, 1961; Clark and Wallace, 1960). This has been variously named basal granule or basal corpuscle in ciliated epithelium, kinetosome in the ciliates, blepharoplast in the flagellates and proximal centriole in metazoan sperm tails. Studies on the behavior and fine structure of this granule reveal a basic similarity irrespective of where it is found and justify its designation by the single term b a s a l   b o d y (Fawcett, 1961). The structural resemblance between the basal body and the centriole is impressive (Gibbons, 1961; Gibbons and Grimstone, 1960; Fawcett, 1961) and more will be said in this connection later. The basal body is not to be confused with a much larger body variously called kinetoplast, kinetonucleus and parabasal body. The term kinetoplast has been used to designate a structure found in the Trypanosomatidae some distance below the base of the reservoir from which the flagellum arises, and gives a positive Feulgen reaction (Clark and Wallace, 1960).

The exceedingly constant architecture from centriole to centriole irrespective of the source deserves some comment. This very fact has been considered on a theoretical basis by Satir and Satir (1964) who present a model for the ninefold symmetry seen in alpha-keratin and cilia, and extrapolate it to include centrioles. Their model for generating a ninefold symmetry is based upon the alpha helix. By assuming that the amino acid in the fifth position of the sequence in the alpha helix is a repeat of the amino acid in the first position, and that the ninth amino acid is also a repeat of the first one, etc., when such an alpha helix is viewed end on, these repeating amino acids would divide the circular cross-section into nine equal sectors. It was then suggested that the already established molecular symmetry would generate the much grosser structure of the cilia and centrioles, too, by fundamentally similar processes. From their considerations one can imagine that the symmetry generating helix would be represented by the hub of the cartwheel frequently seen in the core of the centrioles, and that this need not extend the length of the centriole; its presence at one end of the centriole would be adequate to determine the symmetry and configuration of the entire centriole.

Because there appears to exist an intimate structural and functional interrelationship between centrioles, basal bodies and their derivatives, the scope of this article will not be restricted to the discussion of centrioles.

It is not always possible to decide unequivocally the manner in which particular authors have used the term centrosome and central body; therefore their terminology will be employed followed by (centriole?) if this author thinks that reference is actually being made to centrioles.

B. Distribution of centrioles and basal bodies

Centrioles and structures known or believed to originate directly (some basal bodies, Sotelo and Trujillo-Cenoz, 1958) or indirectly (other basal bodies, trichystosomes and trichitosomes, Lwoff, 1950) from them are very widespread throughout the Kingdoms Plantae, Animalia and Protista. Centrosomes will be discussed when it appears that the investigator may acutally have been describing centrioles and also on the basis of the intimate relationship that exists between centrioles and centrosomes.

The inescapable fact is that, in a great number of organisms, centrioles have not been observed. This is especially true of the higher plants. Now that electron microscopy has provided us with a new criterion for the identification of centrioles, it remains to be seen whether they can be found in those organisms in which the classical cytological techniques failed to reveal them. We are for the moment concerned with the centriole in morphological terms as revealed by light and electron microscopy. It is not to be confused with the term "center", as used by Mazia (1961: p. 117) and others, which is the functional designation applied to a body defining a mitotic pole, whether it is visible or not. Thus, c e n t r i o l e and c e n t e r may identify the same structure, but the former term is based upon morphological criteria and the latter upon functional criteria and it remains to be demonstrated that the functional manifestations displayed by the center can be equated to a morphological unit identifiable as a centriole. For this reason a brief discussion of the distribution of centrioles and related structures is pertinent.

Sharp (1921) concludes that there is no adequate evidence for the existence of centrosomes (centrioles?) in the cells of angiosperms, but they are clearly present in many fungi, algae, certain bryophytes *(Marchantia)* and pteridophytes *(Marsilia)*. Lepper (1956), too, points out the dearth of evidence for centrosomes (centrioles?) among the angiosperms, and cites a few papers describing the presence of centrosomes in the higher plants, but considers them to be in error. Swingle (1897) illustrates centrosomes (centrioles?) during mitosis of the apical cells of the alga *Stypocaulon.* As they migrate toward opposite poles of the nucleus, they are clearly shown to be paired. In *Fucus,* Strasburger (1897) shows the four centrosomes (centrioles?) as mere points in a line on the nuclear membrane. During mitosis these pair off and move toward opposite poles. It appears that at the point of contact of the centrosomes with the nuclear membrane the latter has become distorted. Sharp (1921) and Lepper (1956) cite additional specific examples for the presence of centrosomes (centrioles?) in the above mentioned groups of plants.

Centrioles have been seen in representatives of many animal phyla and very likely occur in all. They are also found in the Protozoa. Just a few specific examples will be used to illustrate the ubiquitous distribution of animal centrioles: Nematoda [Boveri, 1888, 1890 — in cleaving *Ascaris* eggs]; Rhynchocoela [Coe 1899 — in cleaving *Cerebratulus* eggs]; Platyhelminthes [Costello, 1960 — cleaving *Polychoerus* (Acoela) eggs; Minouchi, 1936 — cleaving *Polystomum* (Trematoda) eggs]; Mollusca [Gall, 1961 — spermatogenesis in *Viviparus* (Gastropoda); Conklin, 1904 — first cleavage in *Crepidula;* Galtsoff and Philpott, 1960 — in the oyster *Crassostrea virginica*]; Echinoderms [Harris, 1962 a, 1962 b — *Strongylocentrotus purpuratus* embroys; Boveri, 1900 — in meiotic and first cleavage spindles of *Echinus microtuberculatus;* Wilson and Mathews, 1895 and Hertwig, 1906 — during fertilization in association with spermhead in *Toxopneustes*]; Arthropoda [Kawamura, 1955 — spermatocyte of the grasshopper *Acrydium japonicus;* Johnson, 1931 — male germ cells of the cricket *Oecanthus nigricornis;* Huettner, 1933 — cleavage in *Drosophila melanogaster*]; Chordata [Schreiner and Schreiner, 1905 — spermatocytes in the hagfish

Myxine glutinosa; HENNEGUY, 1891 — in spindles of the germinal disc in trout; RIS, 1955 — in Whitefish blastomeres; POLLISTER, 1933 — nongranular leucocytes and epithelial tissue cells in amphibians; NAGANO, 1959, 1962 — in rooster spermatocytes; BURGOS and FAWCETT, 1955, 1956 — in *Felix domestica* and *Bufo arenarium* spermatids; AMANO, 1957 — in the mitotic spindle of mouse lymph node cells; CONKLIN, 1905 — centrosomes (centrioles?) in the first cleavage of *Cynthia partita* (Ascidian) egg]; Protozoa [CLEVELAND, 1957 a, 1957 b, 1960 a — the appearance, behavior and life cycles of centrioles in flagellates].

The distribution of blepharoplasts (basal bodies) among plants has also been discussed by SHARP (1921) and can be summarized as follows: they have been found in *Oedogonium, Cladophora,* and *Vaucheria* of the thallophytes; in *Marsilia* and *Equisetum* of the pteridophytes; in *Zamia* and *Ginkgo* of the Gymnosperm. In the course of describing spermatogenesis in bryophytes (i. e., *Porella, Astrella, Marchantia, Fegatella, Blasia,* and *Mnium*) WOODBURN (1911, 1913, 1915) discusses the origin and development of the blepharoplasts (basal bodies). This raises the question: do all plants having basal bodies also have mitotic centrioles? It is currently not possible to formulate a definite answer to this interesting question.

The intracellular localization of centrioles is usually highly predictable and can serve as one of the constellation of criteria for their identification. Usually the number of fullsized centrioles present in any single cell is one, two, sometimes three, or four, depending upon the stage in the generation cycle of the cell and the cell type. When one or two fullsized centrioles only are present in a cell (for example, early in mitosis), they may be associated each with a small sister centriole in the act of being generated. An interesting modification of this occurs during atypical sperm formation in Viviparid snails where a parent centriole may be surrounded by a ring containing up to nine radially-oriented procentrioles (GALL, 1961). Each procentriole is the precursor of a daughter centriole. Another interesting deviation from the commonly occurring number of centrioles has been observed by BESSIS et al (1958). In the course of their examination of human (normal and leukemic) hemopoetic organs and tissues from different laboratory animals, they observed megacaryocytes (polyploid cells) containing a large number of centrioles. It was not rare to see 3-6 in the same section. The important aspect is that, for any given cell type, the number centrioles is usually small, discrete and intimately correlated with the stage in the generation cycle of the cell. In interphase cells, on the basis of a large number of cell types, a single centriole, a pair of centrioles or sometimes two pairs of centrioles may occupy a position in the cytoplasm close to the nucleus. The cytoplasm immediately surrounding the centrioles is usually devoid of mitochondria and larger components of the endoplasmic reticulum; and is frequently bounded by elements of the Golgi complex. Early in mitosis the centrioles move a w a y from each other, either singly when only two are present, or in pairs when there are four. This can also be observed in meiosis and frequently in the course of the transformation of a spermatid into a mature spermatozoon. An exception to this has been ob-

served by Sotelo and Trujillo-Cenoz (1958). After completion of mitosis
in neural epithelium of chick embryos (sixty to one hundred sixty eight
hours old) the paired centrioles of the cell migrate t o g e t h e r to the cell
periphery where one comes into terminal contact with the cell membrane
and initiates the development of the cilium while the other centriole remains
very close by. In many cells in mitosis, whenever the presence of centrioles can be demonstrated they occur at the spindle poles. However, this
relationship is by no means universal. There are cells (found in *Pales,
Cypris, Asplanchna, Pisciola* and *Acanthocystis,* see *Dietz,* 1962) in which
centrioles are not normally associated with the spindle poles (both mitotic
and meiotic) or may be experimentally dissociated from them. This can
be seen by the fact that the asters, each organized around a centriole, will
display varying degrees of association with the spindle poles. In *Pales*
spermatocytes, where the asters normally participate in the formation of
the spindle, it can be experimentally demonstrated that a normal spindle
can form in the absence of any asters associated with the poles (Dietz,
1959, 1962). See page 95 for further details. It is unusual to find illustrations of centrioles in which they are not surrounded by an aster (Heidenhain, 1907; Bělař, 1929; Harris, 1962 b). The reverse situation is not un-
common: that is, to find illustrations of asters lacking centrioles or a concentrated centrosome (Boveri, 1890 — in *Sagitta, Ciona* and *Echinus;*
Östergren, Koopmans and Reitalu, 1953 — in plants; Lima-de-Faria, 1958 —
in plants). In these cases one cannot exclude the possibility that centrioles
actually were present, but the techniques employed did not permit the investigator to visualize them.

Cytasters formed in sea urchin eggs artificially activated by a modification of Loeb's "double method" were shown by electron microscopy to
contain at least one centriole (Dirksen, 1961 a and 1961 b).

Although centrioles usually occur in the cytoplasm, in some cases they
are intranuclear. Brauer (1893) describes the intranuclear location of the
central body (centriole?) during the early phases of mitotic divisions in
Ascaris megalocephala univalens testis. By metaphase the centrosomes,
each with a centriole (?), escape the nucleus. Wilson (1928) points out that
the occurrence of intranuclear division centers have also been described for
the öocytes of the copepod *Canthocamptus* and of the platode *Thysano-
zoön.* Presumably these division centers are organized around centrioles.

Basal bodies exist in intimate and specifically oriented association with
the bases of cilia and flagella (Doflein and Reichenow, 1927; Fawcett, 1961;
Gibbons, 1961; Gibbons and Grimstone, 1960, Sotelo and Trujillo-Cenoz,
1958). They represent a constant feature of the structurally complex ciliary
and flagellar apparatus whenever these structures are seen in electronmicrographs of adequate resolution.

When all the organisms in which the cytology of spermatogenesis has
been described are considered (Nath, 1956), one realizes the complexity and
diversity of this phenomenon. Centrioles appear to be present in all
flagellate and non-flagellate sperm. The latter may or may not elaborate
an axial filament which is non-vibratile when present. In flagellated sperm

one usually finds the proximal centriole occupying a position just behind the nucleus at the anterior end of the midpiece; the distal centriole may be immediately adjacent to it or separated from it by the length of the midpiece which can vary considerably. In mammalian spermatozoa (Fawcett, 1958) the posterior end of the midpiece is delimited by the annulus (also called e n d   r i n g   or   r i n g - c e n t r i o l e). This is believed to develop from the larger of two bodies formed by the unequal division of the distal centriole. However, the fine structure of the annulus is so different from that of a centriole that the classical concept of its origin from the distal centriole may not be valid (Fawcett, 1958). Nagano (1962) concludes from his observations that there is no evidence that the ring-centriole originates from the distal centriole. On the other hand, Gatenby (1961) contends that it is not satisfactory to conclude that the ring-centriole is not a true centriole.

Electron micrographs of adult rabbit photoreceptors (de Robertis, 1960) reveal two centrioles within the inner segment, at the base of the connecting cilium. The latter structure contains filaments, arranged in a pattern identical to that observed for cilia and flagella. These filaments pass a short distance into the outer segment of the photoreceptor before disappearing from view. This has led to the conclusion that the outer segment is a modified cilium. Eakin and Westfall (1962) have studied the fine structure of photoreceptors in the ocelli of the hyrdomedusan *Polyorchis penicillatus*. These were found to contain a distal and proximal centriole at the base of a cilium-like structure. This cilium-like structure seems to resemble a cilium more strongly than the connecting cilium of the rabbit photoreceptors. The frontal organ of tadpoles, the pineal organs of lampreys, lizard epiphysis, adult epiphysis and frontal organs in the frog *Rana pipiens* have all been shown to possess cells which closely resemble in their fine and gross structure the known photoreceptive rods and cones of lateral eye retinas (Kelly, 1962 — a review).

Carasso (1958) has studied the ultrastructure of the retinal cells of amphibian larvae (*Rana temporaria*) and has pictures showing flagella "growing" out of centrioles. If a pair of centrioles is shown, only one appears to be associated with a flagellum.

Basal bodies are always found near the cell surface and are a component of the intricate ciliary and flagellar apparatus of cells bearing these organelles (see above under Terminology and General Structure). At certain stages in some ciliate protozoa the kinetosomes (basal bodies) will be present but there will be no cilia. However, when cilia do appear, it is always in association with these bodies (Lwoff, 1950, 1949).

From the above summary it can be seen that there is probably no phylogentic significance to the distribution pattern of centrioles and basal bodies among the animals. Since a basal body always seems to be associated with each cilium or flagellum it can be found in any animal possessing these structures. Even among the Arthropoda, which as a phylum can be considered to lack cilia and flagella, any flagellum observed is associated with a basal body. Since animal and plant asters usually contain a centriole at their center, the construction of an amphiaster by a cell strongly suggests the existence of at least one centriole an each spindle pole. On the other hand, the absence of asters in anastral divsion figures (typical of higher plants and frequently observed in animal meiosis and occasionally mitosis), usually is accompanied by the lack of cytologically demonstrable centrioles. There are exceptions to this. HEIDENHAIN (1907) illustrates centriolar behavior during the division of a red blood cell of a duck embryo and there are no astral rays to be seen at any time during mitosis, although the spindle fibers and centrioles are clearly visible; and HARRIS (1962 a) has observed mitotic spindles in later cleavage stages of sea urchin embryos that possessed well defined centrioles but had no asters. BĚLAŘ (1927) presents an illustrated sequence of both meiotic divisions in the grasshopper *Chorthippus lineatus* spermatocytes in which small asters are visible from prophase to early metaphase of both meiotic divisions, but disappear later in metaphase. The general absence of centrioles and basal bodies from angiosperm cells and their presence in the lower plant groups probably reflects the lack of asters, cilia, and flagella in the higher plants rather than a phylogenetic trend. But there is a qualification to this statement which should be pointed out. LIMA-DE-FARIA (1958) has observed asters during meiosis in an apomictic strain of *Poa alpina,* but no centrioles could be seen. Asters have been observed during mitosis in normal and in aminopyrin treated *Allium* cells (ÖSTERGREN, et al. 1953). These asters were not seen to contain centrosomes (centrioles?).

Thus the presence or absence of centrioles and basal bodies would seem to be correlated specifically with the existence of other structures rather than the phylogenetic position of the organism.

In concluding this section, attention should be called to the reported occurrence of centrioles in yeasts and a bacterium, and the suggested centriolar origin of the myofibrils of striated muscle. LINDEGREN et al (1956), on the basis of light microscopic observations on Giemsa-stained *Saccharomyces,* describe a structure they call a centriole. It appears to occupy the expected position relative to the spindle and chromosomes. MUNDKUR (1954) too, has seen structures identified as centrioles. They are evident in hematoxylin-eosin stained cells and are Feulgen negative. Regardless of the degree of ploidy (haploid, di-, tri-, tetraploid), never more than two centrioles per cell were seen, and occurred singly or in pairs in the cytoplasm in association with the nucleus. A structure corresponding to the centriole described by MUNDKUR (1954) was not observed by HASHIMOTO et al. (1959). In *Bacillus megaterium* two minute Feulgen-positive granules, claimed to be centrioles, were observed by DE LAMATER (1951). WOLBACH (1928) has made observations on histogenesis of myofibrils in tumors of striated muscle origin that led him to conclude that the myofibrils of striated muscle may be of centriolar origin. He identified the granules he observed as centrioles, on the basis of size, morphology and staining properties. It was stated that mitochondria were not stained by the MALLORY's phosphotungstic acid-hematoxylin stain used and therefore presumably could not be confused with centrioles, Line drawing interpretations of accompanying photo-

graphs (poor) show short fibrils extending equidistantly in opposite directions from the granules which he terms centrioles. These observations have
apparently not been verified by independent workers.

Observations by HAY (1962) on the formation of new muscle cells from
mesenchymal cells of the blastema in amphibian regenerating limb may
provide another interpretation for WOLBACH's observations. In an electron-
micrograph of a developing muscle cell of a 17-day regenerate, there can be
seen myofilaments attached to and extending in opposite directions from a
structure identified as a Z-band. There are many such Z-bands with attached
myofilaments in a single cell. These Z-band-myofilament complexes are
reminiscent of the granules with their attached myofibrils described by
WOLBACH (1928). In the event that these two invesitagtors have described
the same structural complex in different tissues, it is possible that what
WOLBACH tentatively identified as centrioles were actually Z-bands.

C. The origin and continuity of centrioles

The question of the origin of the centriole has attracted discussion and
experimental attention since its discovery. As a result, it has generally been
held that centrioles are permanent, self-perpetuating structures. Their behavior has been traced through the life histories of many cell types by
microscopic methods (POLLISTER, 1933: HUETTNER, 1933; WILSON 1898; CLEVE-
LAND, 1957 a, 1961, 1962; HEIDENHAIN, 1907). HUETTNER, HEIDENHAIN and CLEVE-
LAND have shown in their material that the centrioles were visible at any
stage in the life cycle of the cell. This view has received suport from
electron microscopy (BESSIS, BRETON-GORIUS, and THIERY, 1958). If centrioles
are to be considered as self-perpetuating bodies they must be encumbered
with the same restriction basic to the generalized scheme of self-perpetuation
employed by chromosomes. Namely, the production of a "new" (daughter)
centriole requires the presence of a pre-existing "old" (parent) centriole.
Should the "old" centriole of a cell be lost, destroyed or rendered non-
functional before it has directed the production of a "new" functionally
active centriole, the cell would from then on be incapable of producing
another centriole. There are many fine illustrations showing what has been
interepreted as a daughter centriole being generated under the influence
of the parent centriole (SCHREINER and SCHREINER, 1905; JOHNSON, 1931; GALL,
1961; BERNHARD and DE HARVEN, 1960, and others). See Figure 2. From these
illustrations, it is evident that the spatial orientation between the parent
and daughter centrioles is very different from that for sister strands of the
duplicated chromosome. Whereas the sister strands of the duplicated
chromosomes lie beside each other, the daughter centriole is usually oriented
perpendicularly to the long axis of the parent centriole from the moment
it first can be detected until it has become a fullsized centriole. See Figure 3.
This suggests that the presence of the parent centriole may be required only
at the inception of the daughter centriole and can be dispensed with for the
subsequent growth of the incipient daughter centriole into a fullsized centriole. More will be said about this later.

The continuity of centrioles from one cell generation to the next has been unequivocally demonstrated in some excellent studies. By means of diagrams and good photographs, HUETTNER (1933) clearly shows the movements of the centrioles in one complete mitotic cycle in cleavage figures of *Drosophila melanogaster*. The observations demonstrated that the centrioles were visible at all stages of the mitotic cycle with, apparently, no variation in morphology; and the new centrioles always appeared immediately adjacent to the existing centriole and never at random in the cytoplasm. An intracellular structure need not maintain an invariable morphology during the course of its life history, nor even be visible at all times to be considered as a self-perpetuating body (i. e. chromosomes). HUETTNER also observed that the division of the centrioles always preceded the division of the asters. During the ninth, tenth and eleventh cleavages, the primordial germ cells are segregated from the egg and the centrioles of these cells are indistinguishable from those of the cleaving nuclei. He, thus, has offered good evidence for the uninterrupted descent of centrioles through cleavage to the cell lineage which will culminate in gamete formation. In the female *Drosophila* the continuity of centrioles is interrupted at meiosis (which occurs after insemination), for those divisions appear to be anastral. A spindle, but no asters, centrosomes, or centrioles are visible. In view of this the egg presumably obtains the centrioles necessary for the first cleavage from the sperm. Therefore the continuity of centrioles from the parent organism to the offspring would appear to be maintained through paternal "inheritance". POLLISTER (1933) has described centriolar behavior during mitosis in non-granular leucocytes of *Amphiuma tridactylum*. This is a completely different cell type in that it is neither embryonic nor ancestoral to the gametes. Condensation of the chromatin occurred in advance of any change in position of the paired centrioles. Later in prophase the centrioles moved to opposite sides of the nucleus. At metaphase the centrioles were located in the orthodox position and were slightly enlarged when compared to their size observed at early prophase. During anaphase the centrioles were slightly smaller than at metaphase, but for the rest they appeared unchanged. He also reports on the presence and appearance of centrioles in non-dividing cells. Connective tissue, smooth muscle, blood cells (except red blood cells, because the entire cell was stained by the technique he used to discern centrioles in other cells), and epithelial cells all contained a pair of centrioles per cell, usually situated very close to the nucleus. He concludes that they are self-perpetuating bodies characterized more by their behavior and their specific location in each cell type than on the basis of their staining reaction.

In N e r e i s (WILSON, 1898; Fig. 71), the sperm penetrates the cell before the germinal vesicle has disappeared. But soon after it enters the egg, the germinal vesicle breaks down and concomitantly the division figure for the first meiotic division forms adjacent to the disintegrating germinal vesicle. There is one centriole per pole. By early anaphase there are a pair of centrioles per pole, and the sperm nucleus preceded by its minute spermaster (containing a pair of centrioles) advances toward the division figure. Prior to the formation of the second polar body the sperm aster has divided.

After completion of meiosis, when conjugation of the germ nuclei occurs, the egg centrioles and asters have disappeared and the first cleavage amphiaster seems to be organized entirely under the influence of the sperm centrioles. Daughter centrosomes (centrioles?) have been followed through every stage of the first cleavage division into the blastomeres of the two-cell stage in the *Thalassema* egg (Griffin, 1896). A centriole is visible at the focus of the spermaster which divides to form an amphiaster soon after the sperm penetrated the egg. This amphiaster becomes the division figure for the first cleavage division. By late metaphase the centriole at each pole divides into two. During anaphase they migrate to the outer perimeter (furthest removed from the metaphase plate) of each centrosphere and there organize a minute amphiaster in anticipation of the second cleavage division before the first cleavage has taken place. This is a clear demonstration of the direct descent of the second cleavage centrioles from the centrioles of the first cleavage division. The direct descent of the centrioles through the succeeding cleavages was not followed, but it was believed that the continuity was not interrupted. Coe (1899) illustrates a first cleavage anaphase (late) in *Cerebratulus* in which a small amphiaster can be seen at the outer periphery of each centrosphere. The axis of this amphiaster is perpendicular to the axis of the first cleavage spindle.

Further support for the continuity of centrioles can be found in evidence of a different nature provided by Boveri (1896). In studies on dispermic eggs at first cleavage, he observed that in some cells all of the chromosomes went to one pole. Thus one blastomere contained a centriole, but no chromatin. This blastomere continued to cleave and form amphiasters showing that the centriole continued to divide even in absence of chromatin. This observation as well as others already mentioned and still to be discussed strongly supports the concept of the genetic continuity of the centriole.

Conklin (1905) has provided us with a fine description of cell lineage in the egg of the ascidian *Cynthia (Styela) partita*. This study has shown that there are no asters and centrosomes (centrioles?) associated with the meiotic spindle of the egg. The origin of the first cleavage centrosomes is in the sperm which penetrates the cell before the completion of meiosis. The centriole introduced by the sperm, and the aster that develops in association with it, are visible at all stages from sperm penetration to its incorporation into the first cleavage mitotic apparatus. In dispermic *C. partita* eggs, two spindles are formed, usually nearly parallel to each other, which are always structurally independent of one another. The poles of the two division figures are never united into a triaster or connected with fibers to form a tetraster. Consequently, one spindle is haploid and the other ist diploid.

Centrioles are frequently not observed in echinoderm material, but Boveri (1900) has detected them in *Echinus microtuberculatus* and described their behavior. His illustrations show the presence of centrioles and asters in the second meiotic spindle of the öocyte. During early prophase of the first cleavage division the daughter centrosomes, although widely separated, are not yet at opposite poles and each contains only one centriole. Somewhat later in prophase (when the daughter centrosomes are at opposite poles of the nucleus, the nuclear membrane is still intact, and chromosome condensation is occurring,) one can see two centrioles per centrosome. Anaphase is accompanied by a separation of the paired centrioles in each centrosome until they are quite widely separated at late anaphase. No minute amphiaster associated with the centrioles could be detected in either centrosphere.

In WILSON (1897, 1898) there is additional discussion of the wealth of evidence pointing to the spermatic origin of the first cleavage amphiaster centrioles in many different eggs. Earlier BOVERI (1895) had already thought that the egg had no centriole and this led him to develop the theory that the essential event of fertilization was the introduction into the egg cytoplasm of the centriole by the spermatozoon. WILSON (1898; p. 144) arrives at the conclusion that: "During cleavage the cytoplasm of the blastomeres is derived from that of the egg, the centrosomes from the spermatozoon, while the nuclei (chromatin) are equally derived from both germ cells".

It has thus been unambiguously demonstrated in many forms that the centrioles of the first cleavage amphiaster are directly descendant from the centrioles imported by the sperm. On the other hand, observations on some other material have clearly excluded the sperm centriole as being ancestoral to those of the first cleavage division figure. In *Polystomum* (MINOUCHI, 1936) only the egg centriole is involved. LILLIE (1897), working with *Unio complanata*, observed that only the egg centrosomes (centrioles?) enter into the formation of the cleavage amphiaster. A conspicuous comet-like aster containing a minute centrosome appears in connection with the sperm nucleus. This centrosome divides and forms an amphiaster which entirely disappears by late anaphase of the first cleavage. The sperm centrosomes never regain their function. WILSON (1928; p. 442) cites some additional observations which indicate that the centrioles of the first cleavage amphiaster in some animals are derived from the egg centriole. Another conclusion has been reached by CONKLIN (1904) from his observations in *Crepidula*. This is that under normal conditions b o t h the sperm and the egg contribute to the two cleavage-centers. He observed a cleavage centrosome (centriole?) in connection with each pronucleus which persisted until karyogamy. FOL (1891) has insisted on the presence of a centriole in the egg.

Pertinent to the present discussion is a summary by RAVEN (1958) of the development of the sperm aster following sperm penetration in the molluscs. He presents five possible alternative fates for the sperm aster.

1. The sperm aster divides and forms an amphiaster with a central spindle which becomes the cleavage spindle.

2. The sperm aster divides more or less distinctly to form a dicentric aster or two separate asters without a central spindle. These disappear, however, before first cleavage prophase.

3. The sperm aster does not divide, but persists and is involved in the first cleavage spindle.

4. The sperm aster does not divide and disappears before first cleavage prophase.

5. There is no sperm aster at all.

An aberrant case which does not fit any of the above categories is discussed by RAVEN in connection with *Limnaea stagnalis*. The sperm aster appears before the formation of the first polar body and grows rapidly in size during the formation of the second maturation spindle. When the second maturation spindle has become oriented perpendicular to the egg surface, the sperm aster fuses with its inner end to become the deep maturation aster. Following the formation of the second polar body this aster disappears completely before prophase of the first cleavage division.

The continuity of centrioles has been impressively demonstrated also in a number of flagellates by CLEVELAND (1957 a — a review). In a discussion of the different types and life cycles of centrioles in the flagellates, CLEVELAND (1957 a) states that two centrioles, an "old" one and a "new" one, are always present in the resting cell just after cytokinesis. Commencing at prophase two new ones and two old ones are present. The new centriole is produced by the anterior end of an old centriole and frequently remains attached to the latter until it has grown to full size. A new centriole is never seen to appear at random in the cytoplasm.

It is pertinent here to mention observations bearing upon the origin and genetic continuity of structures thought to be closely related both structurally and functionally to centrioles. It has already been mentioned that SOTELO and TRUJILLO-CENOZ (1958 b) have shown by electron microscopy that mitotc centrioles can migrate to the cell periphery upon completion of mitosis in neural tissue and initiate the development of a cilium, thereby assuming the role of a basal body. Thus the same body can at different times function either as a centriole or as a basal body. Similar events in spermatogenesis have been described by GALL (1961) and BURGOS and FAWCETT (1955). Apparently the roles of determining spindle poles and initiation of flagellar development are not mutually exclusive. During the course of spermatogenesis in apyrene sperm of *Pygaera bucephala* MEVES (1903) illustrates an example of this. At the first meiotic division there are two V shaped central bodies (centrioles?) per spindle pole. A fine filament from each arm of the V extends through the cytoplasm and for some distance beyond the cell membrane. The two arms of each V shaped central body separate in the late anaphase and begin to migrate toward what will become the poles for the second meiotic division figure. At the second meiotic division each pole has one central body with its filament. After completion of the second meiotic division each cell has one filament which becomes the flagellum. DIETZ (1962) has observed the very same thing in *Pales*. HENNEGUY (1897) reports seeing flagella attached to centrosomes (centrioles?) of the mitotic figure in an insect spermatocyte. This observation has been often repeated (SHARP, 1921).

Structures identified as kinetosomes are very likely identical with basal bodies (FAWCETT, 1961) and therefore warrant a brief discussion at this point. LWOFF (1949, 1950) gives a general presentation of problems in the morphogenesis of ciliates and flagellates based largely upon observations made by him and his coworkers. He considers in detail the transformations undergone by kinetosomes and their derivatives during the entire life cycle of a number of protozoa. His careful observations revealed that a kinetosome appears always to be formed by the division of a pre-existing kinetosome, which led to the conclusion that it is a cytoplasmic unit endowed with genetic continuity. Its presence has also been detected in non-motile stages of the life cycle. Thus, the uninterrupted descent of the kinetosomes in the course of an entire life cycle has been well established for a number of ciliates. Neglecting the mouth, in an average ciliate (of the *Leucophrys* type) the cilia are arranged in 29 longitudinally oriented somatic rows or kineties (Lwoff, 1949). These kineties are complex structures made up of a fiber, the kinetodesma, and a single (usually) line of kinetosomes. The kinetosomes are always lined up to the left of the kinetodesma. To the left of kinety #1 is the mouth with its membranelles. Prior to division the ciliate increases in size. As this occurs the kinetodesma elongate and the kinetosomes multiply. The transverse division of the ciliate cuts the kineties into two equal parts. Thus the kinetosomes of the equatorial region become either the posterior kinetosomes of the proter (anterior daughter ciliate) or the anterior kinetosomes of the opisthe (posterior daughter ciliate). The proter inherits the old parental mouth while the opisthe has to form a new one. The kinetosomes are also able to divide longitudinally. This is necessary to account for the elongation of the kineties and also transversely in order to produce the oral field where the mouth of the opisthe will appear. In *Gymnodinioides* each trichocyst is generated by one granule which arose from the division of a kinetosome. Thus, under certain circumstances, a kinetosome will divide to produce a granule which in turn will form a trichocyst instead of a cilium. A granule never produces both a cilium and a trichocyst, but only one or the other. Which of these two structures a granule produces depends upon the intracellular environment within which it is located. In *Gymnodinioides* trichocysts are formed only at the tomite phase of the life cycle. *Foettingeria* presents a more involved situation and provides an example of the multiple potencies of kinetosomes when the entire life cycle is considered. The kinetosome is able to divide and produce cilia. In the tomite phase the kinetosome produces a differentiated granule, the trichocystosome, which gives rise to a trichocyst. The kinetosomes of the two-week old trophont can produce another type of differentiated granule, termed the trichitosome, which multiplies and produces trichites. It is stated that the fate of the granule is controlled not only by its position in the organism but also by the phase of the life cycle. A cilia-bearing kinetosome may divide at the end of tomitogenesis and the daughter granule can produce a trichocyst or a trichite. However, once a trichocystosome has expressed its prospective potency by formation of a trichocyst, it is no longer endowed with self-reproducibility. The kinetosome adjacent to the kinetodesma always retains the ability to perpetuate itself irrespective of the prospective fate of its offspring.

Thus Lwoff is of the opinion that one kinetosome is always formed by the division of another, i.e. no d e n o v o formation. Grimstone (1961) points out there is no direct evidence for this; only inferences from fixed material. He argues that LWOFF has not established that kinetosomes divide, but admits that new ones are not formed in the absence of old ones.

The foregoing discussion has dealt primarily with evidence consistent with the origin of centrioles from pre-existing microscopically-detectable centrioles. Thus, centrioles can fulfil the requirement of self-reproducibility essential for bodies endowed with genetic continuity. It is not known if they can meet the second requirement which has to do with mutation (PONTECORVO, 1958). The discovery of cytasters in activated sea urchin eggs by MORGAN (1896) initiated speculation and controversy regarding the possible d e n o v o origin of centrioles. Cytasters (astrosphaeres) can be induced in both fertilized and unfertilized *Sphaerechinus* eggs by exposure for 1 to 2 hours to 1.5% NaCl sea water, but the radial organization of the cytasters in the unfertilized material was much less well defined than of those in the fertilized eggs. When eggs containing cytasters were returned to normal sea water, the cytasters soon faded away. He cites this as indicating that the cytasters are not products of real centrosomes (centrioles?) although they often contain definite central bodies.

After artificial parthenogenesis a bipolar cleavage amphiaster will frequently appear in connection with the nucleus, in addition to numerous cytasters scattered throughout the cytoplasm (WILSON, 1928). The cytasters often have no structural connection to the cleavage amphiaster. He observes that the cytasters at first are scattered throughout the cytoplasm, but later tend to move to the cell periphery and there frequently divide into two. Sectioned material of these stages revealed definite central bodies (centrioles?) in the cytasters and also that the division of the latter is preceded by the doubling of the central bodies. On the basis of their behavior, it would seem that the central bodies are true division centers and arose d e n o v o in the cytoplasm.

Another apparent case of the d e n o v o origin of centrioles has been described by MEAD (1898) for *Chaetopterus pergamentaceus*. After the egg has accumulated much yolk and attained its full size, there appear up to 75 distinct asters (secondary asters) before the onset of meiosis. Then two of the asters (primary asters) begin to predominate in size over the others and continue to grow, while the secondary asters gradually regress and disappear. The nuclear membrane then becomes distinct and a well-defined centrosome (centriole?) appears at the center of each aster. (MEAD uses centrosome to refer to the minute granule at the very center of the aster). These are the centrosomes of the first maturation division. The egg centrosomes seem to disappear after the second maturation division, which makes it highly improbable that they participate in the formation of the cleavage amphiaster.

The question of the d e n o v o origin of centrioles has also been studied by more direct experimental methods and more modern techniques. Invertebrate eggs have been a favorite material for these investigations. One direct approach is to extirpate the centrioles, activate the egg and observe whether or not centrioles reappear. Another technique is to cut the egg into nucleated and enucleated fragments, activate the latter fragment and observe for the development of cytasters with centrioles. The first method was used by McClendon (1908) in observations of segmentation in *Asterias* eggs deprived of chromatin. By means of micropipettes the first polar spindle, or the second polar spindle and the first polar body, were removed. Parthenogenesis was induced by exposing the treated cells to carbonated sea water. The eggs, deprived of both division centers and their chromatin, developed numerous cytasters. The cells segmented into many smaller cells, each of which contained one or more cytasters. No chromatin or nuclei could be detected in the experimental cells. Although there was no mention of centrosomes (centrioles?) it is apparent that the cytasters were functionally indistinguishable from normal cleavage asters. Using the second aforementioned technique Yatsu (1905) described the formation of cytasters and centrioles in enucleated fragments of *Cerebratulus* eggs. When the eggs were discharged into sea water, the first polar spindle formed, but meiosis stopped at metaphase I and did not resume until the sperm had penetrated the cell. While development was arrested at the first meiotic metaphase the eggs were readily cut into two fragments so that the first polar spindle was retained entirely within one fragment and the other was devoid of all nuclear material and centrioles. The enucleated fragments where then treated with a parthenogenetic agent ($CaCl_2$ in sea water). These developed a varying number of cytasters, functionally and structurally like those observed in intact eggs. The cytasters possess typical astral rays and a large central region within which appear a group of dark staining structures which are identified as multiplied centrioles (pluricorpuscular center). In sections of enucleated egg fragments containing cytasters, centrioles were present in most of the cytasters observed (Yatsu, 1904). Obviously the centrioles of the cytasters could not have arisen by the division of pre-existing, microscopically-detectable centrioles for these were completely absent from the enucleated fragments. From other experiments by Yatsu (1908) it appeared that the egg must achieve a certain level of ripeness prior to the operation before cytasters would appear in the enucleated fragment after artificial parthenogenesis. Cytasters would develop in enucleated fragments of *Cerebratulus* eggs only if the germinal vesicle had broken down before the operation. Wilson (1928; p. 690) feels that this adequately demonstrates that "ripening" of the egg is the result of material which is liberated into the cytoplasm when the germinal vesicle disintegrates, and whose presence in the cytoplasm of the enucleated fragment is required for the development of cytasters.

Enucleation experiments of a different nature by Lorch (1952) on sea urchin embryos yielded results which may, in part, be consistent with the view that centrioles can arise d e n o v o. One cell of a two-celled embryo served as the experimental cell, while the other was used as the control. Three situations were investigated. In the first, both the nucleus and the centrosphere (the cytoplasm surrounding the nucleus) were micrurgically removed. No asters or cleavage furrow of any form appeared, but 8 hours after treatment the blastomere divided into a large number of spheroidal cells. The second situation involved removing only the nucleus. Asters appeared and multiplied for up to 7 hours following surgery and then disappeared. By 8 hours this blastomere also divided into a large number of spheroidal cells. The third situation, which was achieved in only a few cells, investigated the consequences of removing only the centrosphere and leaving the nucleus intact. Astral rays always eventually reformed. Neither the nucleus nor the cytoplasm divided before the reappearance of asters. Subsequent to the reconstruction of asters, cleavage appeared to be normal. LORCH feels that FRY's (1929, 1932) results are sufficiently compelling to assume that there are no centrioles in her material and does not discuss the significance of her results in relation to centriole behavior. She does conclude that almost any part of the cytoplasm has the capacity to participate in the formation of cytasters and a normal amphiaster. If, on the other hand, one does assume the presence of centrioles in this material (centrioles have been demonstrated in sea urchin material by BOVERI [1900] and DIRKSEN [1961 a, 1961 b]) and that there is at least one at the center of each aster, it is clear that they cannot arise d e n o v o in cells from which both the nucleus and centrosphere have been removed. The opposite conclusion can be reached if one considers only the results of the second and third situation mentioned above. In these cases either the nucleus or the centrosphere was removed, yet asters appeared in both situations. However, one could reasonably expect that one of these operations would result in the extirpation of the centrioles with the consequence that no asters should form. These results as a whole do not argue for or against the d e n o v o origin of centrioles.

Echinoderm eggs have been subjected to centrifugal forces ranging from those that are sufficient only to displace cytoplasmic particles to forces strong enough to fragment the egg into nucleated and enucleated fragments. HARVEY (1936, 1940) has broken mature, unfertilized eggs of *Arbacia punctulata* into the white, nucleated fragment and a smaller red, enucleated fragment by centrifugation an 10,000 × g. for 3 to 4 minutes in a 0.95 M sucrose gradient. The enucleated fragment was activated parthenogenetically by exposure to hypertonic sea water for 20 minutes. The activated enucleated fragments developed an astrosphere, the beginning of a cytaster. Later two asters were observed and a typical cleavage plane between these two asters appeared. A spindle was never seen, although one was present in normal cleavage divisions. Complete cleavage ensued and development into a late non-nucleate blastula was frequently seen to occur. The blastula had no blastocoel; and if cilia were produced they were short and irregularly distributed. The first cleavage plane following activation of the enucleated half may bear no relation to the centrifugal axis, but usually was perpendicular (i. e. parallel to the stratified layers of cytoplasmic particles) to it. The nucleated fragment would also develop parthenogenetically, but development continued to form a pluteus similar to that obtained from a fertilized enucleated fragment. Unfortunately, there was no mention of the presence or absence of asters in the artificially activated nucleated fragments. This is pertinent in view of observations by MOORE (1938) on centrifuged sand dollar eggs. The cells were subjected approximately 40,000 × g. for 7 to 10 minutes with the result that they were elongated into a dumbbell shape or a pear shape. The nucleus was at the extreme centripital end. The cells were parthenogenetically activated by LOEB's "double" method. The sand dollars whose eggs were used could be classified into three types on the basis of the cleavage patterns: (1) complete but unequal cleavage, with the rate of cleavage at the centripital end slower than at the centrifugal end; (2) the centripital end showed nuclear division and the formation of multiple asters without cytoplasmic division, but the centrifugal half cleaved regularly; (3) eggs in which the nuclear division did not occur and only the centrifugal end displayed cytasters and regular cytoplasmic division. If the cytasters were organized with reference to specific particles, it was clear that these must be numerous; otherwise it would have been impossible for asters to appear in both halves of the centrifuged cells in (2) above.

Successful artificial parthenogenesis, leading to the formation of numerous asters, implies that fully functional centers were derived from the egg cytoplasm. But the problem of whether or not cytasters form with reference to a true centriole has not yet been clearly resolved in the foregoing discussion. It has been frequently assumed that cytasters do not contain true centrioles. BRACHET (1957) indicates that interest in cytaster formation stems from the fact that these structures may form around almost any cytoplasmic granules, eliminating the need for the presence and involvement of centrioles. CONKLIN (1902) suggested that there may be two kinds of central bodies (centrioles?); (a) "artificial" ones which are formed d e n o v o and (b) "true" central bodies which develop in connection with the egg nucleus. In WILSON's opinion, this is improbable.

Although much work implies that centrioles may arise d e n o v o under certain circumstances, the observations are based upon the ability of the cell to produce asters and the presence of centrioles is only inferred. The important question, therefore, becomes whether cytasters are organized around true centrioles or around some body bearing no structural resemblance to centrioles. Electron microscopic examination of cytasters in parenthenogenetically activated sea urchin eggs (by DIRKSEN 1961 a, 1961 b) has shown the presence of what are interpreted to be centrioles. The fine structure of these bodies was not resolved; but: (1) they occupy the position that would be expected of a centriole, (2) there appear to be never more than two per cytaster, and (3) the general shape conforms to that of a centriole. This centriole-like body appears to be similar in structure to the centrioles seen in the normal cleavage spindle (HARRIS, 1962 a, 1962 b). Associated with the centriole-like body DIRKSEN observed a small body of similar electron density which may correspond to the "satellites" or "pericentriolar" bodies described by BERNHARD and DEHARVEN (1960). The cytoplasm surrounding the centriole is relatively clear and free of large particles. This organization is very similar to that seen in fertilized sea urchin eggs (HARRIS 1962 a).

DIRKSEN's observations fairly well establish the existence of centrioles in cytasters, but the question of their origin remains unresolved. Since it is difficult to imagine the spontaneous origin of bodies otherwise capable

of self-perpetuation from non-specific precursors in the complete absence
of a parent body the apparent d e n o v o origin of the centrioles in
cytasters can be viewed (1) as the result of the reassembly of specific subunits into the definitive centriole, or (2) as the result of development from
presumptive basal granules (Mazia, 1961: p. 126) in the cytoplasm. The second
alternative is reasonable when it is recalled that parthenogenetically activated eggs can develop into embryos bearing cilia. Clearly the basal bodies
of these cilia cannot be derived from the sperm centriole. (We are assuming
for the moment that the unfertilized egg has no organized centrioles of its
own.) It can, thus, be imagined that whatever gives rise to the basal bodies
can under certain circumstances develop into the centrioles of cytasters.
There is good experimental evidence (already discussed) which indicates
that centrioles and basal bodies of cilia and flagella are homologous if not
identical structures. Indeed such a relationship was already formulated in
the remarkable "Henneguy-Lenhossek hypothesis" (Henneguy 1898; von
Lenhossek 1898 — discussed by Wilson 1928 and Heidenhain 1907). The
first alternative mentioned at the beginning of this paragraph finds
precedence in the behavior of chromosomes during the mitotic cycle.
Whereas the chromosomes typically disappear from view and reappear
in synchrony with the division cycle, the centrioles are normaly visible at
all phases of the cycle and disappear from view at only infrequent instances (i. e. upon termination of normal mitosis leading to the formation
of öogonia or after the completion of meiosis in öogenesis). The "disappearance" of the centriole may only reflect its disassembly or dispersal
into a form that cannot be detected by light or electron microscopy, but
which retains its information content intact. This could allow it to reas-
semble itself upon a given signal independent of another organized centriole. A similar mechanism is implied in the suggestion of Dirksen that the
egg contains material of pericentriolar bodies (Bernhard und de Harven,
1960) in a dispersed form and that activation of the egg causes its aggregation and subsequent formation into centrioles. It is frequently observed
that the centrioles are "lost" upon the completion of meiosis in the ovary.
although they survive spermatogenesis. This behavior is difficult to reconcile
with a body that is otherwise permanent and self-reproducing, unless it is
capale of such an autonomous, reversible transformation (dispersed state $\rightleftarrows$
aggregated state). Consequently, the genetic continuity of the centrioles
would not be interrupted by meiosis and the ripe egg would be in posses-
sion of structurally dispersed, but otherwise genetically intact, centrioles.

Since the egg appears to contain centrioles (in some form, at least, e. g.
dispersed or aggregated) it can be imagined that something must prevent
them from spontaneously organizing a cleavage figure which would launch
the egg into parthenogenetic development. Also one must account for the
fact that in normally fertilized eggs of most species, the egg centrioles
remain quiescent and the cleavage amphiaster is organized around the
centriole(s) introduced by the sperm (Briggs and King, 1959). How centriole
activity is controlled and regulated by the cell must be viewed as an important problem that remains to be solved.

Another possible source of centrioles are the kinetochores of chromosomes. Observations by POLLISTER and POLLISTER (1943) on typical and atypical spermatogenesis in the snail, *Viviparus malleatus* (n = 9), lend strong support to the concept that centrioles and kinetochores are functionally homologous. They noted that certain spermatids (atypical) contained an abnormally small amount of chromosomal material and a correspondingly large number of centrioles. These not only could be counted, but their presence was verified by the appearance of supernumerary tail filaments. The increase in the number of centrioles could easily be accounted for by assuming that kinetochores had taken on the role of centrioles. This premise could also explain the observation that most of the chromosomes were not included in the telophase nucleus of the first meiotic division. The true centrioles during atypical spermatogenesis behave in a manner expected of normal centrioles. This, in part, is based upon the observation that in late diakinesis there appear about 36 evenly dispersed chromatids. This is four times the haploid number and comes about through asynapsis and the complete separation of sister chromatids. Apparently four of these chromatids have normal kinetochores for at the conclusion of the second meiotic division each spermatid has one. The remaining 32 *acentric* chromosomes degenerate soon after the second meiotic telophase. The second meiotic division is unequal and the division plane is so oriented that all of the degenerating acentric chromosomes as well as a single normal chromosome are included in the larger cell.

The kinetochores of the *acentric* chromosomes are thought to aggregate around the true centrioles sometime before diakinesis to form a homogeneous structure termed the c e n t r i o s p h e r e. At diakinesis the two centriospheres occupy positions near the nucleus and an aster develops around each. Then the homogeneous appearance of the centriospheres disappears and each assumes the appearance of a tightly packed mass of granules, described by MEVES as mulberry-like. In their size, staining reactions and behavior the individual granules are indistinguishable from centrioles and were considered as such throughout the remainder of atypical spermatogenesis. Apparently each centriosphere gives rise to 9 centrioles. During late telophase of Division I, the number of centrioles in each cluster increases until it is approximately doubled. Each centriole thus doubles at about the same time when the single centriole in the typical primary spermatocyte does. Slightly later the compact clump of centrioles disperses and a small aster appears about each centriole. The centrioles take up individual positions at the cell periphery about the time the spindle remnant disappears. The centrioles, still at the periphery, begin to aggregate into increasingly larger clumps until there are two groups at opposite sides of the cell. Each group becomes the focus of a large aster in anticipation of the second meiotic division. The number of centrioles in the two groups is not equal, and it is evident that the group to be included in the larger spermatid contains more centrioles than the other. The centrioles remain in a line just under the cell surface during the second meiotic telophase and each generates a flagellum. Slightly later each centriole divides into a proximal and a distal portion connected by an axial filament. The Polli-sters present other evidence to support their hypothesis that the centrioles of the supernumerary flagella were derived from kinetochores. They classi-fied the spermatids for four species of viviparid snails on the basis of the number of centrioles and determined the frequency of each class. It ap-pears that while the centrioles of the primary spermatocyte in atypical spermatogenesis are divided equally between the two secondary spermato-cytes, the distribution of the centrioles to the two spermatids at the second meiotic division is not equal. Consequently, the sum of the centrioles in the two spermatids derived from the same secondary spermatocyte must equal 2 n (there are 4 n separate and independent chromatids in the atypical primary spermatocyte). Since sister spermatids could not be individually iden-tified, the frequency of the complementary classes of spermatids whose centriole number totalled 2 n was compared to each other. The agreement in frequency between any two complementary classes is remarkable.

At this time another point should be mentioned. Lima-de-Faria (1958) has indicated that at diakinesis the kinetochore of each dyad is structurally double (i. e. one kinetochore per chromatid), although it behaves as a single entity in the course of a normal first meiotic division. This behavior can account for the appearance of exactly nine centrioles (one true centriole and eight supernumerary centrioles derived from the kinetochores) at each pole when the centrosphere assumes the mulberry configuration (in *Viviparus malleatus*). The implication is that the movement of homo-logous kinetochores to opposite poles occurred in a normal manner although they no longer were attached to the chromatids. Since each homologous kinetochore might be structurally double, being made up of two sister kinetochores, the duplication of the supernumerary centrioles during late telophase I may reflect only the separation of sister units and not a net synthesis. As we have seen, the first meiotic division equally partitions the centrioles between the secondary spermatocytes, but the second meiotic division frequently results in unequal distribution of centrioles to the sister spermatids and implies that at this time the centrioles bear no predeter-mined affinity for either pole.

Swingle (1926) describes and illustrates abortive spermatocytes in *Rana catesbiana*. Meiosis proceeds to metaphase I — anaphase I, at which time it ceases. Then a dot-like centrosome (centriole?) fragments into numerous centrosomal pieces. Each piece sends out fine filaments which attach them-selves to the tetrads that are then literally pulled to pieces. However, the Pollisters (1943) suggest that these "extra centrioles" represent detached kinetochores.

Apparently kinetochores in s i t u may possess the ability to organize small asters. King (1901) reports on observations of maturation in *Bufo lentiginosus* that support this conclusion. In late prophase of the first maturation division each of the 12 pairs of chromosomes assumes a ring configuration, termed the chromatin ring, as the result of the fusion of the ends of the chromosomes in each pair. Associated with each chromatin ring is an aster whose center is close to the chromatin ring. These asters are transient and have entirely disappeared by the time the spindle is formed. Although KING considers these asters to occur in the normal course of maturation, the POLLISTERS (1943) question this on the valid basis that she did not observe a complete sequence of stages, and suggest that she observed an abnormal oocyte whose kinetochores were caught in the act of abandoning the chromosomes to become centrioles. Since each chromatin ring appears to be formed from two homologous chromosomes, it is curious that only one aster was associated with each chromatin ring, and never two.

GALL (1961) reports on a careful electron microscope study on typical and atypical spermatogenesis in *Viviparus* and on the basis of his observations offers another interpretation for the origin of the supernumerary centrioles. The primary spermatocytes of the atypical series at first contain two normal centrioles. Subsequently one end of each becomes surrounded by a cluster of procentrioles which did not arise from kinetochores. Following two aberrant meiotic divisions, the former procentrioles give rise to the basal bodies of the multiflagellated sperm. While the POLLISTERS were able to demonstrate a very close correlation between the number of acentric chromosomes and the number of supernumerary centrioles, GALL finds in his material that the number of flagella is more closely correlated with the observed number of procentrioles than the chromosome number. Although he feels that his evidence does not support the POLLISTER theory, it is also not decisively contrary.

While electron microscopy has yielded a clear picture of the ultrastructure of the centrioles, this cannot be said of kinetochores. On the whole these have appeared featureless, even in sections in which the fine structure of centrioles has been resolved. HARRIS (1962 a) has electron micrographs of kinetochores in which it is possible to discern some fine structure. From these it does not appear that the fine structure of kinetochores will be the same as that observed in the centrioles, but a final decision on this point must await additional evidence.

Kinetochores are clearly self-perpetuating structures which exhibit even less variation in their reproduction than that seen in centrioles. Since both of these structures are involved in spindle organization and chromosome alignment, conceptually there is certainly no objection to viewing kinetochores and centrioles as homologous structures. But one must bear in mind some of the differences between these bodies which may present real obstacles to considering them as homologous to each other. Electronmicrographs of favorable sections clearly show spindle fibers attaching directly to the kinetochores. On the other hand, it is quite unusual to see fibers (either spindle fibers or astral rays) attaching directly to the centriole, although both types of structures may be clearly resolved in the same photograph. GALL (1961) in Figure 16 shows an astral fiber that appears to connect directly with one of the triplet fibers of the centriole. Usually the astral fibers seem to terminate some distance from the centriole, leaving it (and any daughter centrioles, pericentriolar bodies or satellites) at the focus of a small area containing little electron dense material.

Another difficulty, which should be resolved before centrioles and kinetochores can be accepted as homologous structures, is related to the geometry of kinetochore reproduction. With the exception of the protozoa, the daughter centriole being generated by the parent always has its longitudinal axis perpendicular to the longitudinal axis of the parent centriole. They never appear to duplicate one along side of and parallel to the other. It is difficult to image that the geometrical relationship between parent and daughter kinetochores would be the same as that usually observed for centrioles.

A third difficulty can be mentioned here. While asters, each organized presumably under the influence of a centriole, may be attached by fibers to one, two, or more other asters to form multipolar figures, kinetochores

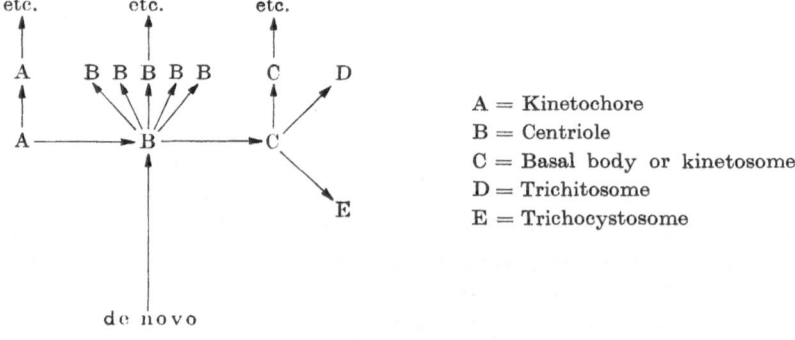

A = Kinetochore
B = Centriole
C = Basal body or kinetosome
D = Trichitosome
E = Trichocystosome

Fig. 7 (see Text)

seem to bear fibrous attachments only to asters and never to each other. Any fibers attached to kinetochores seem always to be directed toward the spindle poles and never directly connect two or more kinetochores to one another. Such kinetochore-to-kinetochore attachments, with the resultant chaos, are strictly prohibited by the LETTRÉS hypothesis.

These are some difficulties which the author feels must be resolved before the centrioles and kinetochores can be considered both structurally and functionally homologous.

During the course of the preceding remarks in connection with the origin and continuity of centrioles, it was pointed out the centrioles may arise from kinetochores and are themselves able to develop into basal bodies. Kinetosomes in some ciliates can generate either trichocystosomes or trichitosomes, in addition to other kinetosomes, depending upon the stage in the life cycle. To what extent are these various bodies interconvertible and can the generation of one type occur under the influence of another type? There is only a partial answer to this question and some of the supporting data has been presented. The diagram in Figure 7 assembles into a single scheme the possible interrelationships that may exist among these various bodies. It should be recognized that this represents a composite of individual steps gleaned from experimental observations on various organisms and is therefore subject to limitations inherent in a hypothetical scheme of this genesis.

The step from kinetochore to centriole is based upon indirect evidence and must be accepted until clearly contrary data become available. This step is actually a functional conversion of A to B rather than the generation of a new structure, B, by A. It is important to recognize that a single kinetochore generates only one sister kinetochore at a time. Before it can generate another one, the sister "must have left home". This is also usually the case for centrioles, but not invariably so. A single centriole (B) has been seen to generate, or rather, to be surrounded by numerous sister centrioles (procentrioles) simultaneously which presumably can develop into full-sized centrioles. The step B to C is based upon firmer experimental evidence. This can represent either a functional t r a n s f o r m a t i o n of a centriole, which no longer determines a spindle pole, into a basal body that produced a flagellum or the actual g e n e r a t i o n of a basal body (or bodies) under the influence of a centriole that still retains its capacity to determine a spindle pole (occurs in the flagellates discussed by CLEVE-LAND). C, of course, is also self-perpetuating. Although experimental evidence exists in support of each indicated step of the hypothetical scheme, evidence bearing on the reversibility of any single step is nearly absent. For example, once a centriole has taken up the position of a basal body and committed itself to that role by producing a flagellum, can it revert and function as a mitotic centriole and/or can it generate another body that can become a mitotic centriole? It is true that during spermatogenesis in some organisms (i. e. *Pales)* each of the four centrioles (two per spindle pole) in the primary spermatocyte has a filament associated with it that becomes the flagellum of the sperm and it is likely that this centriole may also participate in the first cleavage spindle. Only for centrioles is there evidence to argue for a d e n o v o origin; in all the other cases new bodies appear always to be generated by an identical existing structure.

The hypothetical scheme presented in Figure 7 indicates that the inter-conversions are unidirectional. This implies that each is accompanied by a functional differentiation of the body that prohibits it from reverting to the previous or original role. It is not difficult to imagine how this could come about. The kinetochores, centrioles and basal bodies, for example, all occupy very specific and different regions in the cell which makes it entirely conceivable that the specific intracellular environment will modify the prospective potency of the body. Precedence for this concept has been presented by LWOFF (1950) in connection with the behavior of kineto-somes in ciliates. INOUÉ (1959) has stated that basal granules, centrioles and kinetochores may be in fact identical structures, taking on different functions at different loci within the cell.

Some of the main points discussed above concerning the origin and continuity of centrioles are summarized below.

Summary

1. The direct descent of centrioles through successive cell generations has been established beyond doubt for some material, and may very well occur universally.

2. In some material centrioles can be observed during all stages of the life cycle of a cell.

3. There appear to be three possible alternatives to account for the origin of the centrioles of the first cleavage amphiaster.

a) In most cases they are derived from the sperm centriole located in the middle-piece.

b) In a few cases they are clearly derived from the egg centriole.

c) In at least one case *(Crepidula)* the most reasonable interpretation is that one cleavage centriole comes from the sperm and the other is contributed by the egg.

4. Mitotic centrioles have been shown to assume the role of basal bodies.

5. Centrioles have been observed to determine spindle poles and function as a basal body for a flagellum simultaneously.

6. The genetic continuity of kinetosomes is well authenticated. A d d i-t i o n a l l y, kinetosomes upon division can produce differentiated granules which will give rise to structures other than cilia.

7. Centrioles are able to appear in systems known or thought to be devoid of organized centrioles. Such evidence for the d e n o v o origin of centrioles can be accounted for by any of three different interpretations.

a) They arose from non-specific precursors that contain no inherited information.

b) They are the result of the reaggregation of specific dispersed units which bear information.

c) There may exist specific presumptive centrioles (this has been implied: presumptive basal granules — Mazia 1961; pericentriolar bodies or satellites — Bernhard and de Harven 1960). These would contain inherited information which can direct the incorporation of non-specific precursors into a definitive centriole. It should be mentioned that the nucleus is excluded as the source of any specific precursors or presumptive centrioles.

8. There is evidence which implies that centrioles and kinetochores are homologous structures.

Centriole reproduction

1. Evidence from classical cytology, electronmicroscopy and other techniques

Centriole reproduction has been the subject of considerable experimental inquiry. The approach to this problem, utilized by the eminent cytologists of the last century and the beginning of this one, was that of microscopic identification and counting. Then the interest in centrioles and centriole reproduction waned. However, following the period during which interest in these structures fell to a low ebb, they are once again attracting much attention. In part, this is probably the result of their relatively constant substructure as revealed by electron microscopy. But it is also becoming increasingly apparent that centrioles perform a central function in certain cell activities and are indispensible for the organization of some organelles. One of the most important functions in which centrioles are implicated is the internal structural reorganization which a cell undergoes in advance of karyokinesis.

It will be instructive to examine first w h e n centriole duplication occurs, followed by a discussion of h o w it is thought to occur; both on the basis of evidence obtained from light microscopy and electron microscopy. These remarks will not include a consideration of certain flagellate protozoa which will be treated in the next section.

STRASBURGER (1897) described the behavior of centrosomes (centrioles?) during karyokinesis in *Fucus*. These bodies, which very likely were centrioles, appeared as mere points that were intensely stained. He illustrated four centrosomes in a single line on the nuclear membrane. At this stage the nucleoli were still intact and chromosome condensation had not yet commenced. The cells were still in interphase. The centrosomes then paired off and migrated toward opposite poles. During most of this time the centrioles appeared to be in contact with the nuclear membrane and distorted it at these points. Thus, in this case, the centrioles had divided before the visible onset of prophase. A similar situation has been described for mitosis in apical cells in *Stypocaulon* by SWINGLE (1897). In the drawings, centrosomes were illustrated in contact with the nuclear membrane which was distorted by them. As the centrosomes migrated toward opposite poles of the nucleus, they were paired and astral fibers were clearly evident. The spindle was intranuclear and the nuclear membrane disintgrated only as the daughter nuclei were being formed. Before the cell plate was constructed the centrosomes had completed division; there being then four at each pole. Therefore, it would appear that the centrosomes duplicated during the latter phases of mitosis.

BOVERI's (1900) classical observations on the nature of the centrosomes also provide information about when centrioles may duplicate. In *Echinus microtuberculatus,* at early prophase, the daughter centrosomes were widely separated, but not yet at opposite poles of the nucleus; and there was only one centriole visible in each centrosome. Somewhat later in prophase when chromosome condensation was well underway, two centrioles appeared per centrosome. By early anaphase the two centrioles in each centrosome had clearly begun to move away from each other and in late anaphase the centrioles were quite widely separated. In the case of the fertilized *Ascaris megalocephala* ovum a single central body (centriole?) was situated between the pronuclei (BOVERI 1888). It could not be ascertained whether this body was of egg or sperm origin. (During the preceding maturation divisions of the ovum no centrosomes were observed at any time.) The doubling of the central body, in preparation for the first cleave division, occurred at about the same time as the condensation of the chromosomes. The archoplasm showed no radially fibrous organization until after the nuclear membranes of both pronuclei had disappeared from view. By first cleavage anaphase two centrioles per elongated centrosome could be observed.

Other observations in diverse material generally place the time of centriole duplication (i. e. from one per pole to two per pole) sometime during metaphase and anaphase. Some of these observations are briefly discussed in the sentences to follow. HEIDENHAIN (1907) shows that by anaphase there were clearly two centrioles per pole in duck embryo erythrocytes. In connection with a study on the continuity of centrioles during early cleavage in *Drosophila melanogaster* embryos, Huettner (1933) stated "It seems that cleavage of the chromosomes in metaphase and the division of the centrioles occurs simultaneously". In *Cerebratulus*, Coe (1899) observed that duplication of the centrioles occurred during early anaphase of the first cleavage division. From his observations on *Polychoerus carmelensis* eggs, Costello (1961 a) mentions that at metaphase of the first maturation division there was a pair of rod-like centrioles at each spindle pole aligned end-to-end, but that slightly earlier two centrioles could not be resolved. The complete centriole cycle in the goblet cell was followed by Tschassow-nikow (1914), from which it appeared that the centriole duplicated itself by early anaphase or metaphase. The centrioles of the sperm-forming cells reportedly divide during the first meiotic metaphase (Sturdivant 1931). Mead (1898) reported that in *Chaetopterus pergamentaceus* the centrosome (centriole?) at the center of each primary aster had divided by the time the nucleus had disappeared in the course of the first meiotic division, but that during the first cleavage division the centrosomes divided at metaphase. The stage during division when centrioles divide can vary in the same species with the type of division. Thus Griffin (1896) illustrates that in *Thalassema* eggs the centrosomes (centrioles?) have divided by early anaphase of the first meiotic division (which commences only after the sperm has penetrated); after the second meiotic division duplication occurs sometime between early anaphase and telophase (thus there are two centrioles per pole in the second meiotic anaphase stage); and for the first cleavage division the centrioles divided at metaphase. In spermatogenesis of the grasshopper *Chorthippus* Bělař (1927) illustrates the duplication of the centrioles as occurring sometime during the first meiotic metaphase to yield two centrioles per spindle pole. No further duplication of centrioles is shown to occur before the termination of meiosis. Other cases in which the time of centriole duplication varies with the type of division are known.

For additional information on the time of centriole duplication established by classical methods the reader is referred to Wilson (1928), Heiden-hain (1907), and Hertwig (1906).

In connection with h o w centrioles reproduce, much remains unanswered. Classical cytological techniques have been of little value owing to the minuteness of most full sized centrioles. When their reproduction has been described and illustrated, it amounts essentially to "going from one granule to two granules". However, large rod-shaped centrioles do exist in some cells from which clues about their mode of reproduction can be secured. These observations have since been substantiated by electron microscopy of other cells. The spermatocytes of the hagfish, *Myxine*, possess large rod-shaped centrioles whose duplication has been described and illustrated by Schreiner and Schreiner (1905). At the beginning of synapsis a pair of centrioles can be seen lying close to, but not parallel to, each other. During synapsis the centrioles apparently become aligned parallel to each other and on each there can be seen a small bud. The buds were subterminal and oriented perpendicularly to the parent centriole. By prophase they had grown almost to maximum size; and the parent centrioles, each with their nearly full grown daughter centriole, had begun to move toward opposite poles. This pattern of reproduction also was observed in the second meiotic division. During this time the parent and daughter centrioles separated so that they were no longer touching but still close to each other. Each centriole then developed a bud. In some cases an axial filament could be seen which extended from the parent centriole to the exterior of the cell. The daughter centriole had none. It was stated that the axial filament was attached to the opposite end of the parent centriole from that at which the bud or daughter centriole was still attached.

JOHNSON (1931) found that in *Oecanthus nigricornis* the centriole of the spermatogonium was a stubby rod. The two centrioles of the cell were located at opposite ends of the nucleus before the appearance of the spindle. By pachytene of the primary spermatocyte each centriole had become V-shaped, with the apex directed inward and the tip of each limb in contact with the cell membrane. At first meiotic anaphase a split occurred at the apex of the V, resulting in two rod-shaped centrioles. This, too, suggests that the generation of the daughter centriole occurs perpendicular to the parent centriole. Additional illustrations of budding centrioles can be seen in the HEIDENHAIN (1907).

The two preceding paragraphs imply also that centriole reproduction involves a period of growth during which the bud develops into a fullsized centriole.

In contrast to the descriptions in which the axes of the parent and daughter centriole were perpendicular (or nearly so) to each other, there were cases in which the rod-shaped centriole appeared to grow in length and divide transversely into two equal segments. This was described by PAYNE (1927) for spermatogenesis in the hemipteran *Gelastocoris oculatus*. COSTELLO (1961 a) observed that, in *Polychoerus* at the first meiotic metaphase state, the centriole at each pole was distinctly double, consisting of two short rods oriented end-to-end or at slight angle to each other. He considers these to be bivalent. Somewhat earlier in metaphase the centriole at each pole appeared as a single, slightly curved rod (about 1 micron long) with no suggestion of being double. However, owing to their small size, COSTELLO could not be certain that the centrioles which appeared to be a single rod were not also made up of two shorter rods placed end-to-end.

This concludes the classical aspects of centriole duplication. However, before presenting newer evidence bearing on this problem, it will be necessary to say a few words about the hypothetical considerations of self-perpetuating bodies. We associate the property of self-replication of a body with the presence of DNA or RNA in the structure. From many observations, it is perfectly clear that new centrioles are produced from, or under the direct influence of, an existing old centriole. It is evident from the above and from what will be said later that, on the whole, centriolar reproduction does not conform to the usual image of self-replication best described by the "fission-model" (mentioned by MAZIA, et al, 1960) in which

3*

a single body divides into two equal daughter units concomitant with or immediately following its material duplication. The original parent body ceases to exist as such following fission, and each daughter contains both old and new material. Centriole duplication would appear to correspond with the "generative-model" (MAZIA, et al, 1960). This postulates the presence of a "seed" (PENROSE 1959) of molecular dimensions in the parent centriole, which replicates itself. This in turn directs the generation of a daughter centriole which would be a replica of the parent centriole. Phage reproduction is by the generative model. Centrioles are clearly self-perpetuating bodies which very likely contain DNA or RNA and thus possess genetic information. However, we can imagine that this genetic information serves the cell in a very much more restricted manner than the genetic information born on the chromosomes. The information would be necessary only to direct the self-perpetuation of the centriole (i.e. it corresponds to the "seed") and not other cellular activity. The appearance of cytasters in enucleated cytoplasm suggests very strongly that the nucleus does not direct centriole duplication. Centrioles are autonomous units. The role the centriole plays in cellular reorganization prior to division would be a property of the f u l l s i z e d, m a t u r e centriole and not directly attributable to any genetic information it may contain. Thus the only function, albeit important, of any genetic information which may be contained in the centriole would be restricted to directing the production of a new centriole. This would make it very difficult to examine centrioles for mutability, a property deemed essential for a self-reproducing particle having genetic expression (PONTECORVO 1958). Any modification of the information content could interfere with the genetic expression in the form of a new centriole and would thereby prevent the genetic analysis of the parent centriole by depriving us of the daughter centriole.

Newer techniques using the electron microscope have improved our understanding of some aspects of centriole reproduction. In electron-micrographs of favorable sections, two centrioles can frequently be seen at a spindle pole. However, of great importance is the fact that often one of these is much shorter than the other. It has been repeatedly observed (consult references given in section A) that the shorter centriole usually occupies a position at one end of the large centriole and is oriented perpendicular to the long axis of the latter. This consistent spatial relationship indicates that the shorter centriole is very likely the daughter being generated by the fullsized parent centriole, if the inferences drawn from the electronmicrographs are correct. The reproduction of centrioles would seem, on the whole, to be more consistent with the generative model for duplication of selfperpetuating bodies than the fission model. Indeed, it is difficult to imagine how the fission model could be applied to centriole reproduction when both the spatial relationship and the structure of the body is considered. Possible exceptions may be found in the cases already mentioned, *Polychoerus* and *Gelastocoris,* for which the proper spatial relationship between the two centrioles following duplication is consistent with the fission model concept.

GALL (1961) has studied centriole duplication during typical spermatogenesis in the snail *Viviparus* (with electron microscopy). In zygotene and pachytene stages of typical spermatogenesis each centriole is accompanied by a smaller structure oriented at right angles, which is the precursor (or p r o c e n t r i o l e) to the daughter centriole (see Fig. 2). The procentriole is similar to the fullsized or mature centriole except in length. Its diameter is only slightly less than that of the mature centriole and in an "end-on" view it could be mistaken for a mature centriole. It typically does not touch the mature centriole, being separated from it by about 70 mμ. The procentriole is always located at one end of the mature centriole. Its transformation into a mature centriole has not been followed in typical spermatogenesis, since later stages were rare in the material he used. This is believed to occur by elongation of the procentriole with no change in its orientation relative to the parent centriole. From his studies on atypical spermatogenesis, GALL concludes that, in general, centriole multiplication is similar to that in the typical series except in the number of procentrioles produced. In the very early atypical spermatocyte a pair of centrioles can be found in or near the Golgi region. These centrioles are indistinguishable from those found in the spermatogonia and spermatocytes of the typical series. At a later stage the mature centrioles have moved away from each other and in a transverse section it is seen that each has become surrounded by an annulus of radially oriented procentrioles to form the mulberry configuration described by MEVES (1903). Although the details of the fine structure of the procentrioles were obscure, there was nothing to indicate that they would be significantly different from the structure of the procentrioles in the typical series. Examination of a section of a mulberry configuration cut longitudinally through the mature centriole reveals that the procentrioles were not restricted to a single plane perpendicular to the parent centriole. Serial sections suggest that the procentrioles were radially oriented in a spherical array about an imaginary focus situated very near one end of the parent centriole—very much like a cap over the end of a pipe. The total number of procentrioles which may surround one end of a mature centriole was estimated to be between 15 and 20, based upon counts of incomplete series of serial sections. Procentrioles were never seen beyond the immediate proximity of the mature centriole. The details of the subsequent fate of the procentrioles were not determined in this study by GALL, although in broad outline this was established by MEVES (1903) and POLLISTER and POLLISTER (1943).

GALL (1961) draws attention to another interesting aspect of centrioles. The two ends of a centriole differ both structurally and functionally from each other, i.e., the centriole is polarized, in at least four respects: (1) one end produces the flagellum; (2) the other end is associated with the daughter procentriole; and (3) represents the original procentriole or, at least, is derived from it; (4) the two ends also differ owing to the asymmetrical arrangement of the triplet fibers. The flagellum "grows" from one end of a centriole or basal body, defining that as the distal end. In *Viviparus* it was not determined whether the end of the centriole producing the procentrioles was the distal or the proximal one, because the flagella were produced after the relative orientation between the parent and daughter centrioles was lost. But in the spermatocytes of some Lepidoptera (Meves 1903) it can be determined that the procentriole is associated with the proximal end of the centriole, for a flagellum can be seen to emanate from the opposite end identifying it as the distal one.

If centriole reproduction basically follows the scheme for the "generative-model" and the procentriole corresponds to or contains the "seed", another interesting question arises. Is the growth of the procentriole into a mature centriole accomplished by the addition of material on the end facing away from the parent centriole, so that the "seed" (or procentriole) does not change its position relative to the parent centriole; or is the material added to the end facing the parent centriole, causing the "seed" to be pushed away from the mature centriole? Techniques are already on hand which may be able to resolve this question. The cart-wheel arrangement of sub-structure seen in favorable cross-sections through centrioles and basal bodies (see Fig. 1 and 6) appears to be restricted to the proximal end of these structures (Gall 1961, Gibbons and Grimstone 1960). In addition, the cart-wheel structure has been seen in procentrioles. If the cart-wheel arrangement identifies the end of the centriole with the "seed", it remains to be shown what oreientation it maintains relative to the parent centriole during growth. One aspect of centriole behavior may provide a clue in support of the first alternative posed earlier in the paragraph. The author has never seen an electron-micrograph (in the literature) showing a daughter centriole itself generating a centriole until it has become dissociated from the parent centriole. It can be imagined that the daughter "seed" is not able to replicate itself until it has severed its association with the parent "seed". (This could correspond to the s p l i t t i n g e v e n t described by Mazia et al (1960) and discussed below.) Such an arrangement would provide the requisite restriction to prevent "staircase" production of centrioles and would favor the view that growth of the procentriole proceeds by adding material on the end facing away from the parent. However, on the basis of Gall's observations (and others, e. g. Bessis et al, 1958) it is perfectly evident that a parent centriole can generate several daughter centrioles simultaneously. This behavior is not easily reconciled with a simple "generative-model" of centriole reproduction.

Current thinking imagines that chloroplasts and mitochondria, as well as centrioles, are reproduced only from identical pre-existing bodies (Sager 1965). Since the evidence for the presence of DNA in mitochondria and chloroplasts is compelling (Sagar 1965 — an excellent general discussion of nonchromosomal DNA), it is not difficult to imagine that centrioles, too, contain DNA, although there is no direct experimental evidence for this. The DNA of these structures is very probably important in their replication. In this connection Sager distinguishes between *patterned growth* and *template replication*. DNA synthesis exemplifies the latter process while the former involves the orderly addition of new components to pre-existing structures, enlarging the original unit in a two- or three-dimensional repeating pattern. In other words, patterned growth cannot be reduced simply to one-dimensional nucleic acid templates, involved in protein synthesis. The appearance of new centrioles, invariably in the vicinity of pre-existing centrioles, is suggestive of patterned growth, and is consistent with the preceding speculation on centriole duplication.

The electron microscope has really not enhanced our understanding of the method by which material is incorporated into the growing centriole.

BESSIS and BRETON-GORIUS (1958) have observed what they call p e r i - c e n t r i o l a r bodies associated with centrioles of mammalian leucocytes. Their electronmicrographs, which show centrioles in cross-section, reveal that at two levels along the length of the centriole there radiate, like spokes of a wheel, short, slender stalks (65–90 mμ long by 25 mμ in diameter) which each terminate in an apparently structureless knob about 65–73 mμ in diameter. Although the position of the structureless terminal knob is about the same as that occupied by a procentriole, here the resemblance ends. The function of the pericentriolar bodies is not known, but the investigators suggest that they may represent attachment points for the radially oriented astral fibers. There is a photograph which does support such a relationship. SZOLLOZI (1964) has taken electronmicrographs showing spindle fibers attaching to satellites (pericentriolar bodies) arrayed around a centriole. DE HARVEN and DUSTIN (1960) have also observed pericentriolar bodies and suggest either that they may correspond to a preliminary stage in centriole duplication or that they may represent spindle material in a contracted form during interphase.

2. Centriole life cycles in certain flagellates

The extensive observations of the behavior and life cycles of centrioles in different flagellate genera by CLEVELAND (e. g. 1956, 1957 a, 1960 a, b, 1961) are best considered in a separate section owing to the complexity and variation in pattern of centriole behavior among the different genera. He has resolved the life cycles of flagellate centrioles into five types which are reviewed in CLEVELAND (1957 a). These will be briefly discussed in the following paragraphs. Only the immediately relevant points will be presented.

Life cycles of the Type 1 scheme are encountered in *Barbulanympha, Holomastigoides, Idionympha, Holomastigotes, Deltotrichonympha* and other genera. In this scheme, as well as in the other four types, the resting (interphase) cells have two centrioles; an old one and a new one. In this particular instance the two centrioles are elongate, of equal length, and each has a centrosome surrounding the distal (or posterior) end. They are indistinguishable from each other and are interconnected at their proximal (or anterior) ends with a fine fiber. Associated with e a c h centriole is a complete set of extranuclear organelles (i. e. flagellated areas, axostyles, parabasals). At the onset of prophase the fiber connecting the two centrioles is lost and immediately each centriole forms at its proximal end a granule which will ultimately develop into another centriole. This granule is called the n e w c e n t r i o l e. Simultaneously with the appearance of the new centriole the distal end of each old, elongate centriole begins to produce astral rays. The astral rays which grow toward each other from opposite centrioles meet and join to form the central spindle. Soon after, the central spindle begins to elongate as the result of the growth of the astral rays of which it is constructed. Concomitant with spindle elongation the new centrioles, which until now were mere granules, grow in length, and a centrosome appears at the distal end of each new centriole while it is still in the process of elongating. No astral rays are produced by the new centriole during the remainder of this cell generation. The emergence of the spindle causes the two sets of extranuclear organelles to be pushed apart and two sets of new ones, each formed under the influence of a new centriole to be produced. These are generated from the slightly enlarged proximal end of each n e w centriole. This enlargement is the growing point, or reproductive region, of the centriole. The new centrioles continue to elongate and at the time of cytokinesis they are about half grown. Growth of the new centriole continues until it can no longer be distinguished from the old one. All through its genesis the new centriole remains attached to the old one.

The Type 2 life cycle has been observed only in the genus *Trichonympha*. In the resting cell there are two centrioles attached to each other by a fine fiber, but in this case one is long (old) and the other is short (new). The new centriole is actually a mere granule which had been produced during the previous generation. With the onset of prophase the short centriole begins to elongate. However, instead of growing parallel to the old elongate centriole as is the case in *Barbulanympha,* it elongates in a direction at 90⁰ from the old centriole. When the new centriole has become as long as the old one, both begin the production of astral rays from their slightly swollen distal ends. When the astral rays from the two centrioles meet and join to form the central spindle, the proximal ends of the centrioles lose their connection to each other and begin to move apart. Shortly afterwards a new centriole grows out from the proximal end of each fullsized centriole, to which it remains connected. However, unlike in *Barbulanympha,* the new centrioles do not grow during the remainder of this cell generation. Soon after it is produced, the new centriole begins to generate a new set of extranuclear organelles and this process continues until the time of cytokinesis. Thus the new centriole forms flagella, parabasals and other extra-nuclear organelles in its first generation (the generation in which it was formed), and only in the next generation does it produce astral rays from the distal end and a daughter centriole from the proximal end. This situation prevails in most flagellates.

The Type 3 centriole life cycle has been observed in the genera *Pseudotrichonympha, Trichomonas, Urinympha, Leptospironympha* and others. In the resting cell both centrioles are short curved rods with the concave sides facing each other and held together by a broad interconnecting strand or bar. These lie at the anterior end of the cell and at prophase both begin to elongate posteriorly at the same time. Before elongation has proceeded very far the centrioles separate and a centrosome begins to develop at the distal end of each. When only slightly longer, the production of astral rays begins. The astral rays, which are growing toward each other from opposite centrosomes, soon join to form the central spindle. Simultaneously with the continued elongation of the centrioles, the central spindle elongates, the centrioles continue to move away from each other and a new centriole appears at the proximal end of each elongating centriole. The new centriole does not grow at all during the remainder of this cell cycle. After cytokinesis the spindle remnants and that portion of the elongated centriole which grew since prophase degenerate and disappear. When this has occurred the old centriole cannot be differentiated from the new one; thus restoring the resting cell condition. As in the other genera, each new centriole begins the production of extranuclear organelles shortly after it has been generated by the proximal end of an old centriole, which means that there are enough new extranuclear organelles to give each daughter cell the same number as the parent cell had before cell reproduction began.

The Type 4 life cycle has been observed only in the genus *Macrospironympha*. In the resting cell both centrioles are elongate and in this sense they resemble the pattern observed in *Barbulanympha*. These centrioles are also connected to each other at their proximal ends by a fine fiber. There are no centrosomes surrounding the distal end of each centriole in the resting cell. Instead there is a single large body, termed the rostral body, surrounding the distal ends of both centrioles. In early prophase, the nucleus, which lies a short distance posterior to the rostrum and to which it is connected by a nuclear sleeve, loses this connection and migrates posteriorly. Shortly following the migration of the nucleus, the rostral body also migrates posteriorly. Although the nucleus takes nothing with it when it migrates, the rostral body carries with it the distal ends of the elongate centrioles, i. e. that portion of the centriole responsible for the production of the achromatic figure. The detached distal ends of the centrioles appear as dots either within or attached to the surface of the rostral body. In this position the detached ends of the centrioles begin to form the achromatic figure. The apparent function of the rostral body was to carry the distal ends of the centrioles to the point where the nucleus and the achromatic figure are to make connection with each other. Shortly after the achromatic figure has come in contact with the nucleus, a centrosome can be seen to form around each dot-like distal end of the centrioles. After nuclear division the two nuclei migrate to daughter rostra (which had developed in the meantime). In the process they leave behind the remnants of the central spindle, the centrosomes and the distal end of each centriole, which all disintegrate and disappear completely. Concomitant with the formation of the central spindle, the separated old centrioles are each producing a new centriole. The new centrioles become full-sized and a rostral body develops around the distal ends of each pair of centrioles by the time nuclear division is completed.

The Type 5 centriole life cycle is seen in *Joenia, Mesojoenia, Rostronympha, Joeninia,* and other genera. In this scheme neither of the two centrioles present in the resting cell is ever elongate in any stage of its life cycle. In shape these centrioles are stubby rods about twice as long as broad, which can be distinguished from each other (except during the early stages of the formation of the new one in prophase) by the fact that the new centriole always has a fine fiber connecting it to the flagellated area and other extranuclear organelles which it produced during the previous generation, while the old one does not possess such a fiber. The centrioles of *Joenia* do not have separate achromatic figure-producing regions and extranuclear organelle-producing regions, as do so many genera of hypermastigotes (Types 1–4). In the resting cell the two centrioles lie fairly close together and are held in this position by a small, early central spindle between them. This central spindle was formed at metaphase during the previous cell generation. As soon as this spindle begins to grow at prophase, it pushes the two centrioles apart and a new centriole appears beside each old one. Each new centriole is connected to its corresponding old centriole by an early central spindle. Therefore, from now until cytokinesis the cell has three spindles, but only one is functioning. In *Joenia* (and other genera with the same type of life cycle) there is only one set of extranuclear organelles present in the resting cell, instead of the two sets observed, for example, in *Barbulanympha* and *Trichonympha*. This comes about when, early in prophase, the nucleus and centrioles become dissociated from the extranuclear organelles, which subsequently degenerate. Therefore no parent axostyles, parabasals, flagella, etc., are carried over during division into the next generation. After division each daughter cell has a new centriole and an old centriole. Only the new centriole produces a set of extranuclear organelles, providing the cell with only one set.

The basic features of the centriole life cycles in flagellates may now be summarized. Two centrioles, an old one and a new one, are always present in the resting cell. Beginning with prophase there are two old centrioles and two new centrioles present. With the exception of the life cycle typical of *Pseudotrichonympha*, the new centriole remains attached to the old one until it has grown to the same size. The proximal end of a centriole produces both the new centriole and extranuclear organelles, but never at the same time. In the life cycles Types 1 through 4, the centriole produces a set of extranuclear organelles in the same cell generation in which it was itself produced, and forms a new centriole only in the following generation. Type 5 life cycle is apparently an exception to this in that the new centriole produces both extranuclear organelles and another centriole in the same cell generation; namely, the generation following the one in which the centriole was produced. The distal ends of the centrioles, in some cases, are surrounded by a centrosome and produce the astral rays and spindle fibers. These are produced in the generation following that in which the centriole itself was produced.

3. Multiplicity of centers and mode of reproduction

In the foregoing pages the morphological aspects of centrioles have been considered and inferences made regarding their functional relationship to pole formation. When centrioles can be resolved at the poles of the mitotic apparatus, it is presumed that they were responsible for polarizing the structure. However, it still remains to be demonstrated that the functional aspects of pole determination can be equated to a structurally identifiable centriole(s). The term "center" is the functional designation of whatever is responsible for polarizing the mitotic apparatus when the structural basis for pole formation cannot be resolved. Under normal circumstances, when centrioles can be seen at the poles of the mitotic apparatus, usually two per pole are observed, although early in mitosis only one per pole is not infrequently seen. Is this duplex character necessary in order to form a single functional pole? Or is e a c h centriole potentially able to determine a pole? Also, what is the m u l t i p l i c i t y of the centers when particulate centrioles cannot be detected? Would it correspond to the number of centrioles at the mitotic poles when these can be seen?

The question of the multiplicity of centers has been investigated by an experimental method that does not depend upon resolving visually any discrete structures. These studies were conducted by Mazia et al (1960) in eggs of sea urchins and sand dollars. In this material centrioles cannot clearly be identified by light microscopy, but the presence of conspicuous asters aids in identifying the centers. The basic observation which led to this study was that cells placed into sea water containing mercaptoethanol (0.075 M) at about the beginning of first cleavage metaphase were blocked at metaphase. However, when they were removed to normal sea water at about the time the control cells were undergoing the second cleavage, they cleaved directly from the one-cell stage into four cells (Mazia and Zimmerman 1958). The subsequent, more detailed experimental analysis provided the solution to this preliminary observation. The cells placed in the mercaptoethanol (ME) were arrested at metaphase and the chromosomes remained in the metaphase configuration throughout the block (approximately one hour in duration). However, the direct cleavage into four cells, termed quadripartition, implied the presence of four mitotic centers. This was confirmed by microscopic examination of the division figure. Yet, at beginning metaphase when the ME block was imposed, there were only two identifiable mitotic centers. (It was also demonstrated that the quadripartition procedure was applicable to the second and more typical division cycle.) Therefore, during the course of ME blockage, the mitotic apparatus changed from the normal bipolar condition to the quadripolar state. The splitting of each mitotic center present at the onset of the ME block, and its separation to form a tetrapolar division figure, can be observed in samples of cells removed at intervals during the course of the block. All during the transformation the chromosomes remained condensed and in the metaphase configuration. It is significant to note that the time course for the appearance of four centers was not greatly affected by the ME. The control embryos at the second cleavage division each have four centers, while the cells blocked by ME at first cleavage metaphase and allowed to remain in the ME sea water until the control embryos were dividing into four cells also had four centers.

The presence of four centers in the ME-treated cells can be accounted for either by: (1) the normal duplication of the centers during the ME block, or (2) the splitting and separation of what were originally duplex centers.

If the second alternative is correct, each of the four daughter cells upon completion of quadripartition would receive only half of the normal complement of center material, and this deficiency could be expected to reflect itself at the next division. This was indeed the case, for the cells entered the next division with only half of a mitotic apparatus; a perfect monopolar figure. It appears, then, that the ME interferes with the duplication of centers, but permits the separation of existing units, and leads to the conclusion that each center is normally duplex, but that it is capable of being split into two functional centers. Under normal conditions each center is propagated as a double entity and its duplication is a process in which two units give rise to four. Additional evidence to support the view that the centers are normally duplex and that the role of the ME is to interfere with their duplication comes from a consideration of the time at which duplication of centers occurs. Since, as we have seen, the normal bipolar division figure contains four potential centers which normally function in pairs, but which can be induced to function individually so that each determines the pole of a tetrapolar mitotic figure, it follows that duplication of centers occurred sometime prior to the ME block. If the cells were placed into ME before the duplication of centers had taken place, then the mitotic apparatus should possess only two potential centers and quadripartition could not occur. This, too, has been observed. Center duplication for the first cleavage division coincides with the completion of pronuclear fusion, while for the second division it is the time of transition from first cleavage telophase into interphase. Mazia et al. (1960) have further inferred from their data that the s p l i t t i n g   e v e n t (defined as the time when sister centers become independent) must occur very shortly after the d u p l i c a t i o n e v e n t (defined as the doubling of the number of potential poles). Thus, their analysis of mitotic center behavior in sea urchins and sand dollars leads them to resolve the overall aspects of reproduction of centers into three processes: (1) the duplication event in which a copy of the original is produced (this may be further resolved into the replication of a part of the center, the "seed", which in turn directs the growth of the complete new center); (2) the splitting event; and (3) the physical separation of the centers following splitting. It is suggested by Mazia et al. (1960) that the "old" units cannot produce "new" units until they have split. This can also be taken to mean that sister centers must split from each other before either one can produce its own likeness. Although this analysis of mitotic center reproduction is based entirely upon functional considerations, without visualization of the structural basis for these manifestations, a close correlation does exist between the inferred behavior of centrioles in other animals determined by the method of "looking-and-counting" and the behavior of mitotic centers in the sea urchin and sand dollar. It has been pointed out earlier that typically there are two centrioles per spindle pole (excepting many meiotic spindles of the second meiotic division). When only one can be seen it occurs early in mitosis. However, of the two centrioles, one may be a small one in the act of being generated under the influence of the other centriole which is always full sized. Sometimes four centrioles can be seen restricted to a small space before the spindle poles have been determined (SCHREINER and SCHREINER 1905, BERNHARD and DE HARVEN 1960). In this case two of the centrioles are "new" each apparently in the act of being generated from a full sized "old" centriole. The statemnet by MAZIA et al. (1960), to the effect that the duplication event must be preceded by the splitting event, finds support from cytological evidence. Normally one does not find a centriole being generated by a daughter centriole that has not begun to move away from its parent centriole. That is, the first generation centriole cannot produce a second generation centriole until the former has become structurally dissociated from its parent centriole.

In *Barbulanympha* CLEVELAND (1958) has shown that when only one pole (distal end of the elongate centriole) is present, no chromosomes were ever connected to it. Hence the chromosomes that should have been attached to this pole never moved. When there are two poles present, but only one is near the nucleus, the poles do not cooperate to form a central spindle and the chromosomes behave just as though there were only one pole—no movement. This is true also when more than two poles are present but with only one near the nucleus. Although the behavior of the poles in *Barbulanympha* is somewhat different from that of the mitotic center in sea urchin eggs, the poles are clearly important in spindle formation.

There is another group of observations which may have an important bearing on centriole behavior. These are concerned with the effect of ultraviolet and X-ray irradiation upon mitosis. HENSHAW and COHEN (1940) and HENSHAW (1940 a–e) have systematically investigated the induction of multipolar spindles by X-ray irradiation of the gametes of *Arbacia punctualata* prior to fertilization. The accessory asters so produced were clearly fibrous and many engaged chromosomes. The exposure threshold for the production of multipolar spindles was definite, and the range between this level and the amount required to induce 100% multipolar cleavage was very narrow. Accompanying multipolar cleavage was a delay in the onset of cleavage. The curves which described the delay in cleavage and the production of multipolar spindles suggested that these were not dependent upon the same change induced by the radiation. It was suggested by HENSHAW that the accessory asters were formed as a result of changes in the nuclear material. By means of a very simple dilution experiment, RUSTAD (1959 c) provided strong support for HENSHAW's contention that the multipolar spindles formed following fertilization with irradiated sperm were not a polyspermic effect. Ultraviolet irradiation can also delay cleavage in *Arbacia* (BLUM and PRICE 1950). However, they employed different experimental designs. In one, eggs were irradiated 10 or 28 minutes after fertilization. When the cells were irradiated 10 minutes after fertilization, the first cleavage was delayed about 50 minutes. The time interval between the first and second cleavages was extended 15 minutes. In control cells the first cleavage normally occurred 53 minutes after fertilization and the second cleavage followed the first by 30 minutes. A different picture emerged from cells irradiated 28 minutes after fertilization. The first cleavage was delayed only 20 minutes, while the second cleavage occurred

60 minutes later; i.e. delayed by 30 minutes. In the other experimental scheme, cells were irradiated at different times following the first cleavage and the degree of delay of the second cleavage determined. The maximum sensitivity to the ultraviolet occurred immediately after cleavage, after which it dropped to a minimum by about ten minutes after cleavage. Exposure to ultraviolet from this time until the second cleavage did not delay the second cleavage, i.e., the cells were insensitive. Thus the interval between the first two cleavages can be divided into two periods with respect to ultraviolet sensitivity: the first period, from cleavage until ten minutes later, was one of changing sensitivity beginning with the maximum at the completion of cleavage and decreasing at an accelerating rate to almost zero by ten minutes after cleavage; this was followed by the second period during which the cells were essentially insensitive to ultraviolet at the same

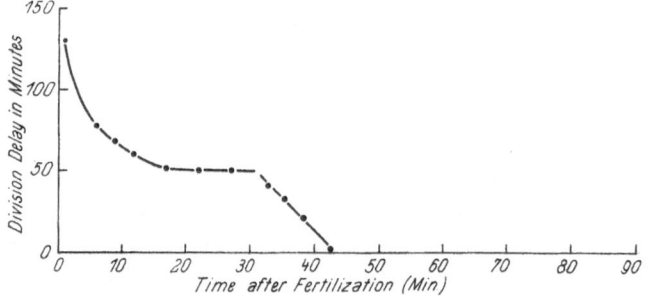

Fig. 8. This curve shows changes in sensitivity to ultraviolet-induced mitotic delay of the first cleavage division in the sea urchin *Strongylocentrotus purpuratus*. The fifty percent division time delays following irradiation of samples with 400 ergs/mm² as a function of when the exposure was administered between fertilization and division (90 min. after fertilization). The cells were kept in the dark following irradiation to prevent photoreactivation which is known to occur in this species. The division cycle shows four phases of UV sensitivity: 1) a period of very high sensitivity rapidly fall to a lower plateau; 2) the plateau during which there are no changes in sensitivity; 3) a linear transition from the plateau to zero sensitivity; 4) period of complete insensitivity to this UV dose. From Rustad, Exp. Cell Res. **21**, 1960, 596—602.

dose. Blum und Price suggest that the delay in cell division following exposure to ultraviolet had its origin in the nucleus.

Rustad (1960) also has investigated the changes in the sensitivity to ultraviolet-induced mitotic delay during the cell division cycle of the egg of the sea urchin *Strongylocentrotus purpuratus*. He observed that the first division cycle progressed through a series of four phases of ultraviolet sensitivity (Fig. 8). The highest sensitivity was exhibited during the first fifteen minutes and was thought to be associated with the early events of fertilization rather than events directly associated with mitosis. The sensitivity decreased rapidly to a plateau which characterized the second ultraviolet-sensitive period. During this period, the sensitivity remained unchanged. The third period was the straight line transition from the plateau to the fourth period of zero sensitivity. The third, or transition, period occurred immediately before the "streak stage" was visible in v i v o. The "streak" appeared when the asters migrated to opposite sides of the nucleus. The division times for the first cleavage of cells irradiated during the insensitive period were indistinguishable from the division time of the control cells (Fig. 9). This was true even when doses eight times greater than the customary one were used. In some of the experiments involving unusually high ultraviolet doses, the cells irradiated during the insensitive periods exhibited no first division delay, but were unable to divide a second time. Changes in sensitivity to ultraviolet-induced mitotic delay during the second mitotic cycles were also studied, because the first division cycle included events peculiar to fertilization that are not part of the mitotic cycle. Three phases were observed: (a) plateau, (b) transition, (c) insen-

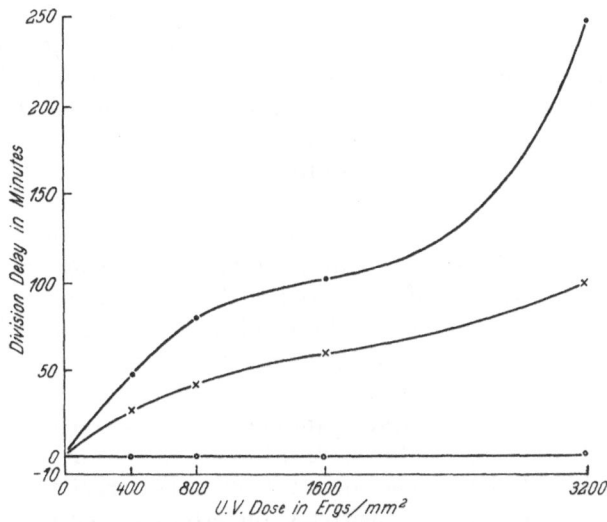

Fig. 9. These curves show fifty per cent division time delays of the first cleavage division as a function of UV dose for samples irradiated before fertilization (crosses), 13 minutes after fertilization (dots) and 75 minutes after fertilization (open circles). Observe that zero time delay is plotted 10 minutes above the axis. The second division is uniformly delayed by UV irradiation given during the insensitive period for first cleavage division delay. From Rustad, Exp. Cell. Res. **21**, 1960, 596—602.

sitivity. The p l a t e a u of the second division began during the insensitive portion of the first cycle. Thus, cells irradiated during the insensitive phase of the first mitotic cycle did not escape damage, but the expression of radiation damage as mitotic delay did not appear until the second mitotic cycle. From the basic pattern of ultraviolet-sensitivity changes (plateau, transition, insensitivity,—which have been observed in other material by GIESE (1949), CARLSON (1940, 1954), RUSTAD infers that a single event, occurring only during the transition phase, was sensitive and that ultraviolet damage incurred earlier was stored without recovery until that period. RUSTAD (1959 a, 1959 c, 1959 e) indicates that the observations on ionizing and ultraviolet radiation induced mitotic delay are consistent with the hypothesis that this arises from damage to a single biochemical process which controls the duplication or separation of the centriole. Observations on multipolar spindle formation by X-ray irradiation of gametes before fertilization also suggests a similar target. Additional experimental evidence was cited to support the centriole hypothesis of radiation-induced mitotic delay. For example, hypotonic sea-water can induce a monastral block when sea urchin eggs were treated before a critical stage in the mitotic cycle. This critical time corresponded approximately to the transition period. Thus, he observed mitotic delay in monastrally-blocked cells that were ultraviolet irradiated during the insensitive phase of the control cells, fertilized at the same time, and then returned to normal sea water. Reciprocally, ultraviolet irradiated cells could be monastrally blocked with hypotonic sea-water long after the control cells had passed the critical stage.

RUSTAD feels that the data of HENSHAW (1940 a—e), HENSHAW and COHEN (1940), and BLUM and PRICE (1950) reflect centriole damage rather than nuclear damage.

Acridine orange (10^{-4} M to 10^{-7} M) has been observed to block or delay mitosis when administered 20—25 minutes after fertilization. When added later, the second division rather than the first was affected (RUSTAD 1959 b, 1961 b). It was postulated that the delay pattern, which was very similar to that observed for ultraviolet irradiation, reflected damage to or interference with the RNA of the centriole.

X-ray irradiation of cultured mammalian cells greatly increases the incidence of multipolar mitoses. LEWIS and MARIN (1963), using guinea pig cells, observed that the frequency of X-ray induced multipolar mitoses was clearly dose dependent, and that the multiplicity of the mitotic poles also increased with the dose. They suggest that the multipolarity did not result from a primary effect of the X-rays upon the duplication of centrioles, but was a secondary consequence of a radiation induced block to division. On the basis of another study of X-ray induced multipolar mitoses in KB human cells, FETNER and PORTER (1965) suggest that the primary target for multipolarity had approximately the same sensitivity (or size) as gene mutations. However, they feel that a single or functionally double centriole does not fulfill the requirement for the target, for they observed neither a sequential appearance of tripolar and tetrapolar spindles, nor a differential frequency in their occurrence. In contrast to this is the excellent correlation between the t r a n s i t i o n p e r i o d (see Fig. 8) in sensitivity to UV-induced division delay, reported by RUSTAD (1960), and the occurrence of the postulated *splitting event* (MAZIA et al., 1960) in the reproductive cycle of mitotic centers. Thus at the moment it is impossible to decide unambiguously what is the target in the discussed UV and X-ray irradiation studies, if it is even the same target.

There are earlier reports in the literature of multipolar cleavage and nuclear division without cytoplasmic division which may provide additional insight into centriolar behavior. LOEB (1892) placed *Arbacia* eggs for varying lengths of time into sea water containing 2% NaCl. No segmentation or cleavage took place in this medium. However, when the cells were returned to normal sea water, they divided about twenty minutes later. Interestingly, the number of "cleavage spheres" into which they divided directly approached the same number they would have divided into through successive cleavages had they remained in normal sea water during that time. These results were also observed by NORMAN (1896) and were not the result of polyspermy. Clearly, the structures, presumably centrioles, responsible for establishing mitotic poles, were able to multiply in the absence of cytokinesis. The very consistent relationship between the duration of exposure and the number of "cleavage spheres" into which the cell divided indicated that the duplication of centers was proceeding more or less normally in the absence of cytokinesis. Therefore, the production of multiple centers was probably not the result of the same series of events that led to the formation of the multiple centers (or, very probably, centrioles) required for the appearance of numerous cytasters in artificially activated eggs. Clearly, center or centriole duplication can be uncoupled from cytokinesis.

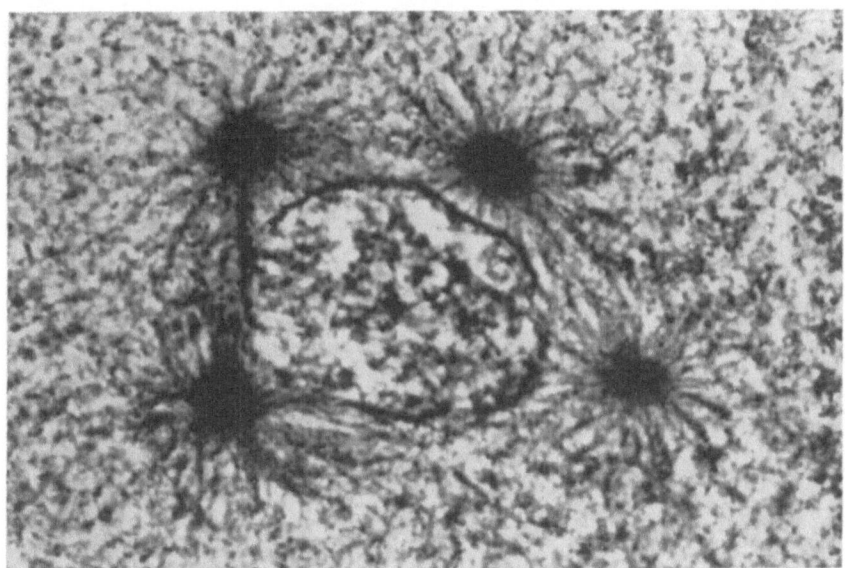

Fig. 10. A multipolar prophase induced by caffeine in a Urodele egg (*Triturus helveticus*). From Sentein (1961 b), Pathologie-Biologie 9, 445—466.

With the aid of phenylurethane, applied at very specific times during the second mitosis in cleaving urodele eggs, Sentein (1961 a, 1961 b, 1962 a, 1962 b) can induce the formation of multiple centers. See Figs. 10 and 11. Other agents such as caffeine (see Fig. 10), and chloral also are effective in producing multipolar figures. These can be induced at about any stage in mitosis. The agents arrest cleavage before mitosis is arrested, but apparently do not interfere with the formation of multiple poles. In some cases (Fig. 11) there are formed more poles than can be accounted for merely by the separation of existing potential poles. In these instances duplication of poles must have occurred. Under proper conditions the multiplication rate of the centers appears to more rapid than normal. Multipolar spindles have also been observed by Barthelmess (1957). But not in all cases of multipolar spindles do we find a centriole for each spindle pole (Dietz 1959, 1962).

Multipolar cleavage induced by 1% urethane has been observed in *Chaetopterus* eggs by Schuel (1961). The experimental cells divided directly into as many cells as seen in the control embryos fertilized at the same time, following a 15–20 minute delay after removal from urethane.

It has been stated by BUCHER and MAZIA (1960) that the earliest known event in the sequence leading to mitosis is the duplication of mitotic centers. Indeed the duplication of centrioles and centers, in preparation for the division during which they are to function as autonomous entities, frequently occurs during the preceding division. This has been adequately discussed elsewhere. The question was then asked by BUCHER and MAZIA whether the duplication of mitotic centers (or centrioles) might serve as the event initiating a sequence of interdependent processes; or whether

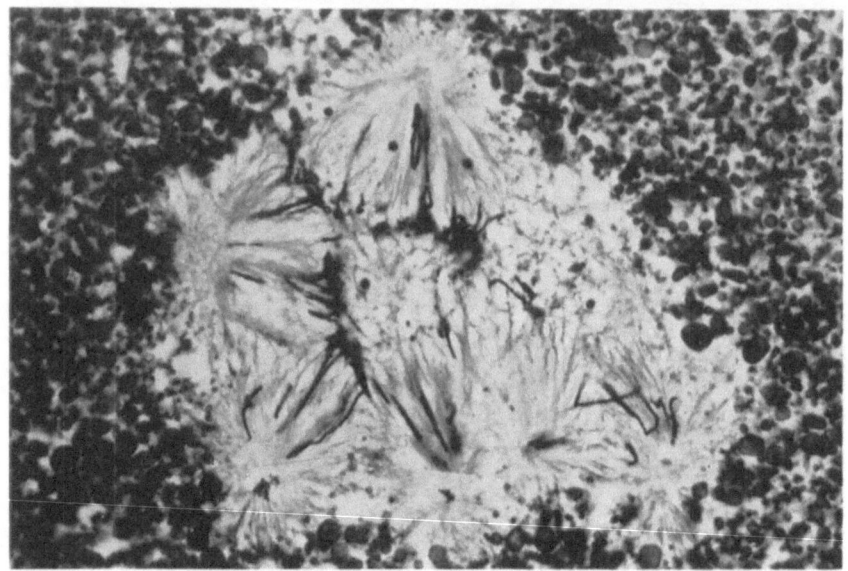

Fig. 11. A multipolar anaphase induced in a Urodele egg by exposure to ¼ saturated phenylurethane. From SENTEIN (1961 b), Pathologie-Biologie **9**, 445—466.

the individual processes occur independently in parallel and the mitotic machinery can go into operation only after all these processes are completed. If the first alternative is valid, then successful preparation for division in advance of the mechanical events of mitosis could be interrupted by failure at any point of the sequence. Thus, for example, DNA synthesis could not occur if the centers had not duplicated. Furthermore, it has been shown that: (1) sea urchin eggs rapidly took up thymidine (BIBRING 1961); (2) mercaptoethanol inhibits mitosis if applied any time prior to metaphase and inhibitis duplication of mitotic centers, but does not interfere with the splitting of the centers (MAZIA et al. 1960); (3) mercaptoethanol as such does not prevent the incorporation of thymidine into DNA (BIBRING 1961); (4) cells blocked at metaphase by mercaptoethanol do not incorporate thymidine into chromosomal DNA as long as the chromosomes retain the metaphase configuration (BIBRING 1961). With this knowledge BUCHER and MAZIA (1960) experimentally inquired into the dependency of DNA synthesis upon center duplication. The essence of the experimental design was to add mercaptoethanol to a suspension of fertilized eggs, either b e f o r e or a f t e r the centers had duplicated. Ten or twenty minutes later tritium-labelled thymidine was added and the suspension was incubated long enough to permit the splitting and separation of any centers present. The delayed addition of thymidine allowed adequate time for the mercapto-ethanol block to become established. The incorporation of tritium-labelled thymidine into DNA was determined by autoradiography. It was clear from the results that DNA synthesis did occur in the absence of prior duplication of the centers. A l l the nuclei were labelled even when the mercaptoethanol was added early enough to block fusion of the pronuclei. However, when the cells were blocked at metaphase following the addition of mercaptoethanol at prophase, n o metaphase stages were labelled.

One can recognize a number of discrete mechanisms capable of exerting some control over normal cell division (MAZIA 1961, SWANN 1957). These occur in a definite temporal and spatial sequence that suggests the presence of machinery to integrate them into the normally observed coordinated pattern. But it is equally apparent that trauma to particular mechanisms will not necessarily prevent other mechanisms from functioning in a normal manner. For example, pronuclear fusion, which normally precedes DNA formation in the sea urchin, can be blocked but DNA formation proceeds at the proper time as determined by the control cells. Other examples could also be cited. Clearly the interdependence of a number of mechanisms responsible for the events preparing the cell for division is more apparent than real.

E. Synthetic activity of centrioles and related bodies

There is a great dearth of information about this important aspect of centrioles. Information about the synthetic pathways leading to the for-mation of discrete centrioles and the time schedule for the incorporation of various non-specific precursors may reveal whether the centrioles are assembled from subunits synthesized in advance or whether the incor-poration of non-specific precursors occurs concomitantly with the appear-ance of the definitive centriole. Such information could resolve whether any of the three alternatives mentioned on Page 32 (The Origin and Continuity of Centrioles) accounts for the d e n o v o origin of centrioles.

It cannot be said with certainty that centrioles contain DNA and/or RNA, although their ability to perpetuate themselves implies that either or both may be present. The cytochemical techniques used to visualize cen-trioles are too nonspecific to give any definitive, positive information regarding their chemical composition.

DIRKSEN (1961 b) studied the uptake of tritiated uridine by developing sea urchin embryos. By means of autoradiography it was demonstrated that no activity appeared at the center or elsewhere in the aster, even after several cell generations. These negative results are not decisive. There are three interpretations offered for the absence of activity in the centriolar region of the aster: (1) the centrioles contain no RNA; (2) the centrioles contain RNA, but are constructed from preexisting subunits which do not incorporate additional RNA before they are aggregated into the centriole;

(3) the centriole contains so few RNA molecules that, owing to the long half-life of tritium, one could not expect to detect any labelled RNA molecules in the relatively short exposure period used (calculations are presented to support this interpretation). Most of the uridine taken up was incorporated in an unidentified labile fraction, which could account for more than $2/3$ of the total label taken into the cell. The chromosomes were found to be strongly labelled, but it was not determined whether the uridine had been incorporated into DNA or RNA.

Kinetosomes have been studied recently for chemical composition and synthetic activity by Seaman (1960, 1962). He has been able to achieve a mass isolation of kinetosomes from *Tetrahymena pyriformis* by the selective solubilization of cell constituents with digitonin, and subsequent isolation of the kinetosomes by ammonium sulfate fractionation. The use of digitonin was suggested by the technique of Child and Mazia (1956) used for the isolation of cilia in large quantities from the same organism. Chemical analysis revealed that the protein and DNA content of kinetosomes was much higher than that found in the whole cell extract, while the lipid and carbohydrate levels were much lower. The enzymatic activity of kinetosome suspensions was also investigated (Seaman 1960). It was observed that the activity in units/mg. protein for glycolysis and oxidative phosphorylation was greater by factors of 4 and 10 respectively than for the whole cell extract. Apyrase, succinic dehydrogenase and fumarase activities for kinetosome suspensions were also greater than for the general extract. It was also shown that these suspensions could bring about an increase in biuret-positive material when incubated in a suitable medium. The incorporation of C^{14} glutamic acid paralleled the rate of increase of biuret-positive material and can therefore be considered an index of synthesis. The incorporation was strongly inhibited by DNAase, RNAase and only slightly by high concentrations of chloramphenicol. The incorporation of glutamine, arginine, histidine, isoleucine, leucine, lysine, phenylalanine and valine into 5% TCA-insoluble material was reported. The activity of a suspension of kinetosomes was different from mitochondrial and microsomal systems. Seaman concludes that the kinetosomes, on the basis of the chemical composition and enzymatic activity, are well equipped to carry out the role of self-duplication and protein synthesis.

F. Centriole orientation in relation to cleavage

The minuteness of most centrioles makes it next to impossible to observe the spatial orientation retained by them, relative to each other, at opposite poles of the spindle. Studies on the few organisms with sufficiently large, rod-shaped centrioles (excluding the Protozoa), has given some insight into this aspect of centriolar behavior. The mutual behavior of centrioles in dividing cells has been observed and discussed by Costello (1961 a). From his work it appears that the orientation of the centrioles determines the position in which the daughter centrioles will separate from each other; and that the position of the main axis of the spindle is determined by the path along which the daughter centrioles separated. The spindle axis, in turn, establishes the positions of the daughter cells relative to each other. The reader is referred to COSTELLO's article for a more detailed discussion.

II. Structure and Formation of the Mitotic Apparatus

The mitotic apparatus is a transient structure present only during the brief interval of the generation cycle when the cell is in the period of division, and it is in a state of ceaseless change. This is readily intelligible when it is assumed that the essential function of the mitotic apparatus is the quantitative and qualitative equipartition of genetic material and information at the time of nuclear division. However, not all cells undergo the fundamental change in gross structure associated with preparing the hereditary material for equipartition and constructing the mitotic apparatus. In procaryotic cells the DNA in each nuclear body seems to be associated with a single structural element and nuclear division is essentially amitotic. The macronucleus, a highly specialized type of nucleus found in eucaryotic cells of ciliates, also divides amitotically (STANIER and VAN NIEL 1962). Amitotic division is feasible in this instance owing apparently to the highly polyploid nature of the macronucleus. However, in most eucaryotic cells the genetic information of the cell is neither contained in a single structural unit nor is it present in a highly polyploid (or redundant) condition; with the consequent result that a mitotic apparatus is usually present during nuclear division (see BUCHER 1959, for exceptions).

MAZIA (1961: p. 236) makes the cogent statement that the mitotic apparatus is both a r e g i o n and a b o d y. When viewing this structure as a body we consider only what is structured and coherent; when viewing it as a region we must also consider small molecules and special physical-chemical conditions (diffusion gradients, local variation in ion distribution) which cannot be discovered under the conditions favorable for elucidating structure. Studies on isolated mitotic apparatus, which are "native" in appearance, disclose that they were arrested in whatever stage of mitosis the cell had entered when it was disrupted and as yet no conditions have been discovered under which the isolated figures will move chromosomes. Thus the question of the unstructured internal environment of the mitotic apparatus looms crucial and may hold the key to the complete understanding of mitotic apparatus function.

HOFFMAN-BERLING (1954 a, 1954 b) has extracted fibroblast cells in mitosis with a water/glycerol solvent system to obtain cellular models analogous to the fiber models prepared from muscles. Chromosome movement of later anaphase stages could be induced in these models by the addition of 1.5×10^{-3} M ATP. The poleward movement of the chromosomes appeared to be the result of a pushing force exerted by the interzonal region (see also BĚLAŘ 1929) rather than a pulling force. When this is done with cell models in early anaphase, the movement of the chromosomes appears partly to be the result of a poleward-directed pulling force exerted between the poles and the chromosomes.

The mitotic apparatus can, in a general way, be thought of as a gel insofar as this term can be used to describe a cohering mass of unspecified macromolecules interacting in ways that permit them varying degrees of orientation (from highly oriented to completely random) in response to specific conditions (Mazia 1961). Thus, the mitotic apparatus can be isolated as a coherent body lacking all of the microscopically detectable fibrous organization normally associated with the structure (Mazia and Zimmerman 1958). In this connection the interesting observation of Harris (1962 a, 1962 b) should be mentioned. Her electron microscope studies showed that in mercaptoethanol blocked cells, the spindle fibers themselves were not destroyed, but only disoriented. Additional discussion of the mitotic apparatus as a physical body, as well as chemical considerations, can be found in Mazia (1961) and below.

There is already much literature extant about the formation, structure and possible mechanism of action of the mitotic apparatus. These include the following reviews and articles more restricted in scope: Wilson (1928), Bělař (1929), Ronnevie (1947), Cornman (1944), Hughes (1952), Schrader (1953), Kupka and Seelich (1948), Wada (1950) and Östergren, Molè-Bajer and Bajer (1960). In Mazia (1961) can be found a provocative and very comprehensive discussion of mitosis and the physiology of cell division to which the reader is referred for the additional information necessary for a complete picture. Because of this wealth of existing information, only the more recent evidence (principally that from electron microscopy, polarized light microscopy, interference microscopy, chemical and cytochemical techniques) will be considered as it applies to the formation and structure of the mitotic apparatus.

A. Evidence from physical methods

Schrader (1953) presents a fine general review of experimental evidence from living material, including birefringence studies, that has led to the concept that the spindle possesses a labile, longitudinally fibrous organization. This has received further support from more refined polarized light microscopy (Bajer 1961, Inoué and Bajer 1961). Schrader and Inoué and Bajer (1961) pointed out that, with the exception of a very few cases (some flagellate protozoa and the mite *Pediculopsis graminum*), spindle fibers cannot be observed in normal living cells by ordinary light microscopy. However, by means of a sensitive polarizing microscope, Inoué (Inoué and Bajer 1961) has observed the microscopic structure and submicroscopic organization of the mitotic spindle in a diverse variety of cells by virtue of the birefringence of the spindle fibers. He has observed chromosomal and spindle fibers during division in living cells of: pollen mother cells; oocytes of marine worms and *Mytilus;* developing eggs of medusae, ctenophores and echinoderms; spermatocytes in *Drosophila* and some orthoptera; and endosperm cells of *Haemanthus katharinae* (Fig. 12). Whenever the material was healthy, the optical conditions were favorable, and an adequately sensitive instrument of suitable resolving power was used, the chromosomal and continuous fibers were seen to be birefringent. In

method (MAZIA 1955). The pressure effected drastic disorganization of linear and radial structure of the spindle-aster complex. The effects were reversible. In general the lower the pressure the longer the exposure time necessary for equivalent degrees of disorganization. Some evidence suggests that the mid-region of an anaphase spindle is especially susceptible to disorganization. Presumably the disorganization of the the mitotic apparatus was the result of pressure-induced solation of intracellular gel structure.

RUSTAD (1959 d) has investigated isolated mitotic apparatus with the interference microscope. When viewed in 1% digitonin, the isolated figures had a maximum optical thickness of 1 wavelength (546 mμ). The chromosomes were optically thicker than any other constituent. After the centro-

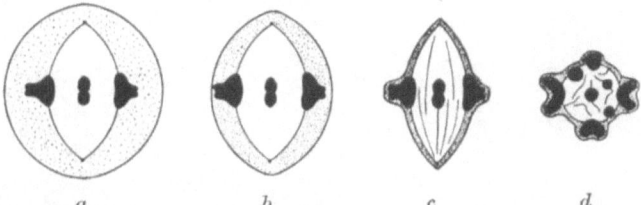

a   b   c   d

Fig. 13. Schematic representation of the action of hypertonic medium on *Chorthippus* spermatocytes in metaphase. *a*) a cell in an isotonic medium; *b*) a slightly shrunken cell; *c*) a strongly shrunken cell; *d*) a polar view of (*c*). The cytoplasm is represented by stippling. The density of the stippling indicates the degree of shrinkage. It can be seen that the change in spindle volume is much less than the change in cytoplasmic volume. The spindle gives the impression of being a semirigid structure. From BĚLAŘ, Arch. EntwMech. Org. **118**, 1929, 359—484.

some appeared as a distinct sphere in phase contrast, it was observed to be considerably less dense than the surrounding astral rays. The metaphase spindle displayed a continuous increase in path length from the outer border to the axis connecting the centers of the asters. In late anaphase the interzonal region increased in mass. This was accompanied by an increase in basophilia of the same region. LONGWELL and MOTA (1960), using interference microscopy on fixed material, noted a similar increase in mass of the interzonal region during anaphase of meiotic divisions in wheat.

Analysis of mitosis in living endosperm cells of *Leucojum aestiveum* L. by interference microscopy was conducted by AMBROSE and BAJER (1960) and the results were compared with earlier studies of birefringence changes during mitosis. Their measurements show that the dry mass per unit area of the spindle region progressively increased with the appearance of birefringence. The dry mass per unit area of the chromosomes was seen to increase steadily during prophase until the second phase of "clear zone" formation. But when the "clear zone" was developing, the dry mass per unit area of the chromosomes decreased considerably, although the chromosomes still contracted somewhat in length and increased slightly in diameter. This would indicate that material was secreted from the chromosomes and perhaps contributed to the formation of the "clear zone" and possibly the spindle. They conclude that the birefringent structures, which appeared during mitosis, were the result of secretion or synthesis leading to localized increases in concentration and were not merely the result of reorientation of molecules already present in the region. This differs somewhat from another opinion already presented above. Clearly, if birefringence is accompanied by a progressive increase in concentration of macromolecular material, reorientation of molecules (or micelles) alone cannot account for the appearance of fibrous structure in the spindle.

The celebrated experiments of Bělař (1929) on spindle elongation in *Chorthippus* spermatocytes reveal a number of interesting facts. His basic experimental design involved placing primary and secondary spermatocytes into more or less strongly hypertonic Ringers solutions at specific stages of the first and second meiotic divisions, and observing the sub-

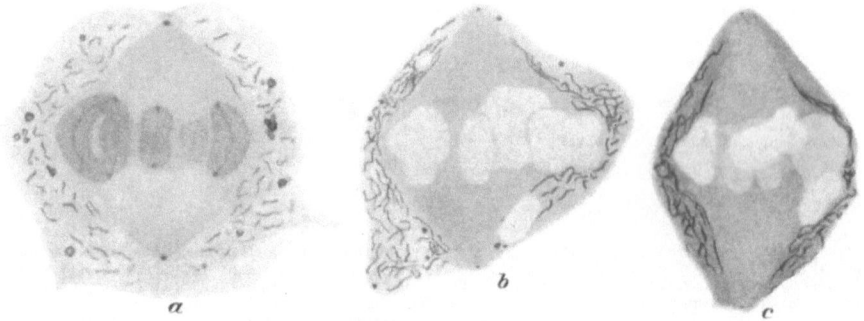

Fig. 14. This figure compares a normal *Chorthippus* spermatocyte with spermatocytes exposed to hypertonic media. *a*) is a normal primary metaphase spermatocyte (stage V); cells (*b*) and (*c*) are primary spermatocytes at stage IV exposed for no longer than three minutes to a Ringer's solution, whose salt concentration was five times normal, before fixation. Cell (*b*) is somewhat shrunken and cell (*c*) is more strongly shrunken. From Bělař, Arch. EntwMech. Org. **118**, 1929, 359—484.

sequent events associated with division. It is very clear that the spindle has considerable rigidity and/or cannot be delimited from the remainder of the cytoplasm by a semipermeable membrane in the usual sense of the word. When cells in metaphase were placed in increasing hypertonic Ringers solutions, the overall shrinkage by the cell was proportionately considerably greater than the volume decrease of the spindle. See Figs. 13 and 14. This clearly demonstrates that either the spindle as a whole does not behave as an osmometer, or it is rigid. Other evidence obtained by Bělař clearly demonstrates that the spindle is a fairly rigid structure. When cells were placed into a strongly hypertonic Ringers solution (the osmolality was 4.5 times normal) at stage IX (late anaphase) of the first meiotic division spindle elongation continued against the opposing force exerted by the shrinking cell. The result was a spindle bent double into a U-shaped configuration (see Fig. 15). When this same experiment was performed using cells at stage XIII (late telophase and the cytoplasm deeply constricted) of the first meiotic division, spindle elongation was much less, resulting in a less severely deformed spindle (see Fig. 15). These data suggest that not only is the spindle a fairly rigid body, but that it is capable of elongating against the relatively considerable opposing force that must be exerted by the shrinking cell. Also the degree of elongation possible under hypertonic conditions decreases as the stage at the time of initial exposure to the hypertonic conditions advances from anaphase to telophase. Spindle elongation may be entirely consistent with the already described increase in dry mass of the interzonal region of some other cell types.

Bělař has proposed that the chromosomes are attached to the equatorial region of the pole-to-pole elements (which he designates Stemmkörper) whose elongation causes their separation. Cleveland (1958) holds a similar opinion in the case of *Barbulanympha*, for he has suggested that the elon-

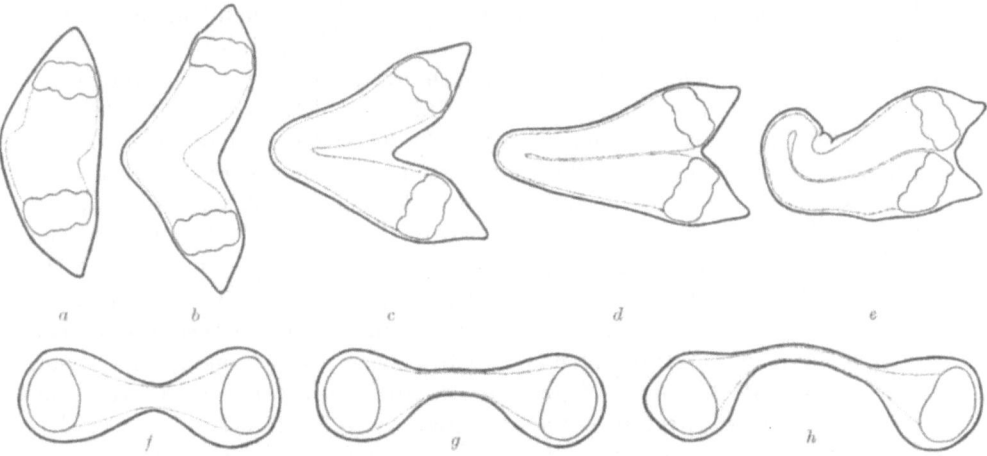

Fig. 15. These drawings show the extent of spindle elongation in living cells that can occur in relation to the stage when the cell was placed into the hypertonic medium (osmolality was 4.5 times that of normal Ringer's). Drawings *a–e* show the sequential changes for one cell placed into the hypertonic medium at stage IX, and *f–h* of another cell that was at stage XIII when placed into the hypertonic medium. Both cells were in the same preparation and simultaneously exposed to the same hypertonic medium. From Bělař, Arch. EntwMech. Org. **118**, 1929, 359—484.

gation of the central spindle fibers places tension on the chromosomal fibers causing the poleward movement of the chromosomes.

The rigidity of the spindle has also been demonstrated by micromanipulation studies of Carlson (1952) and K. Kawamura (1960). Cleveland (1958) observed that the central spindle in *Barbulanympha* behaves as a rigid body even when it is cut away from one of its centers.

B. Evidence from chemical methods
1. Sulfur-containing compounds

This leads to the question of the nature of the crosslinks responsible for maintaining the subunits (micelles or molecules) of the mitotic apparatus in an orderly array. The findings of Rapkine (1931), based on chemical analysis, represent an early attempt to approach the problem of cell division by chemical means. They revealed fluctuations in TCA-soluble sulfhydryl concentration during the first cleavage division in the sea urchin egg. The results were interpreted in terms of converting intramolecular protein disulfide bonds into intermolecular bonds that were presumably related to the build-up of the mitotic apparatus. Rapkine assumed that the TCA-soluble sulfhydryl material he measured was glutathione, but this was subsequently shown not to be the case. By carefully duplicating RAPKINE's experiment, SAKAI and DAN (1959) substantiated his original observations, but found that the TCA-soluble material was a protein or polypeptide. Fluctuations in protein and soluble sulfhydryl which correlated with meiosis and microspore mitosis in both *Lilium* and *Trillium* have been observed by STERN (1956, 1958).

SAKAI (1962 a) found a contractile protein in the KCl-soluble protein fraction extracted from the water-insoluble residue of homogenized sea urchin eggs. This appears as a fibrous precipitate when the extract is stirred in acetone or distilled water and can be formed into a thread. The thread model contracts vigorously under the influence of di-, tri-, and tetra-valent metal ions; this reaction is reversible with EDTA. The model also contracted in the presence of oxidizing agents (cystine, oxidized glutathione, and dehydroscorbic acid) and the contraction could be reversed by reducing agents such as cysteine or ascorbic acid. The amount of work the thread model could perform increased as the load was increased. Further studies (SAKAI 1962 b) revealed that: 1. threads prepared at different stages of the first cleavage division differ in contractility, and 2. the water-soluble fraction of the egg contained an active agent which could cause the thread model to contract. Maximum contractility occurred at metaphase, after which it dropped sharply. Of perhaps greater importance is the relationship of the ratio of protein-bound-SH groups to Protein-N (or the SH/N ratio) to the percentage contraction of the thread. As the SH/N value increased from 1.5 to 2.0 the percentage contraction induced by $CdCl_2$ increased from 31 to 43. When the water-soluble fraction was used to induce contraction, the percentage contraction increased from 4 to 13. The contraction-inducing activity of the water-soluble fraction was associated with the disappearance of SH groups in the thread. Presumably these were oxidized to S-S bonds. The role, if any, of this contractile protein in mitosis is unknown but the conditions most propitious for its contraction occurred at meta-anaphase.

More direct approaches have been employed to elucidate the nature of the crosslinks responsible for maintaining the structural integrity of the mitotic apparatus. One involves cytochemical techniques and another is concerned with the nature of the conditions consistent with the successful isolation of the structurally intact mitotic apparatus from cells (living and preserved) and the properties of these isolated division figures. A careful cytochemical study of sulfhydryl groups, localized in the region of the mitotic apparatus, during the first cleavage in three species of sea urchins and one species of heart urchin were performed by KAWAMURA and DAN (1958). Their results revealed a general increase in stainable protein-SH before metaphase followed by a decline during the later stages of division. The decline was most evident in the centrosphere (or astral center) and the interzonal area. The region between the chromosomes and the centrosphere remained intensely stained until telophase. Apparently different results were obtained by SHIMAMURA, ÔTA and HISHIDA (1957). They stained for both sulfhydryl and disulfide groups in another species of sea urchin and demonstrated localized fluctuations in sulfhydryl and disulfide levels. Their photographs clearly showed that regions stained for sulfhydryl groups frequently complemented the areas of mitotic figures at the same stage stained for disulfide groups. This was most apparent in the anaphase and early telophase stages. In these, the centrospheres stained only lightly for sulfhydryl, but very intensely for disulfide groups. At anaphase the chromo-

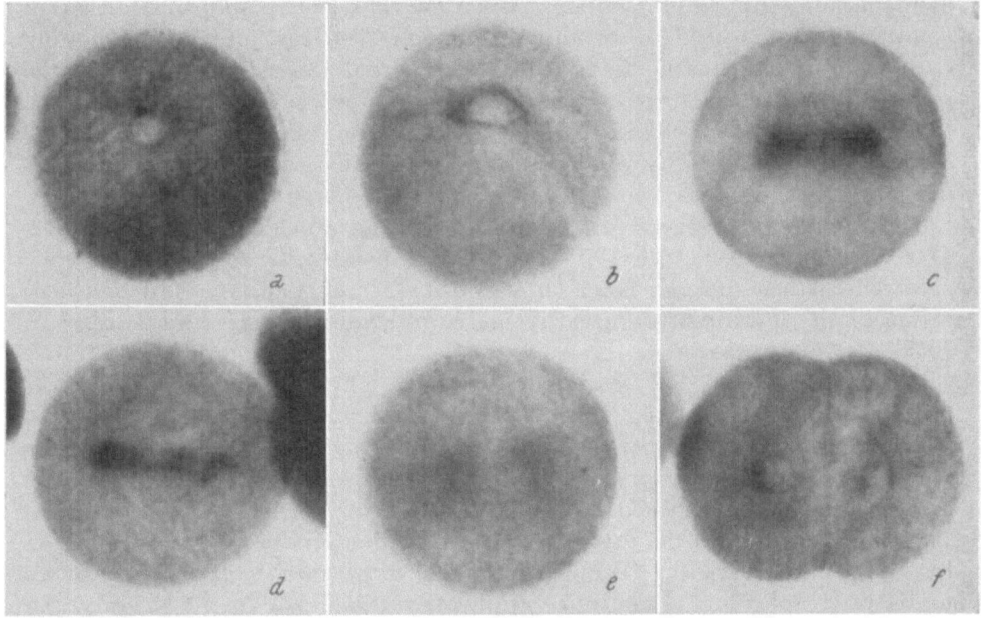

Fig. 16. Eggs of *Arbacia punctulata* fixed in 5% trichloracetic acid stained with Bennett's reagent /1.4-(chloro-mercuriphenylazo)-naphthol-2/. These cells were stained without prior reduction of disulfide groups, and consequently exhibit the distribution of free sulfhydral groups. Cell (*a*) is an unfertilized egg; cell (*b*) is in the streak stage 40 minutes after fertilization; cell (*c*) is metaphase (60 min.); cell (*d*) is anaphase (60 min.); cell (*e*) is telophase (60 min.); cell (*f*) is two cell stage 74 minutes after fertilization. The progressive change in number and distribution of free sulfhydryl groups can clearly be seen in this sequence. From KAWAMURA, Exp. Cell Res. **20**, 1960, 127—138.

somal fibers stained strongly for both groups while the interzonal fibers appeared to stain only moderately for sulfhydryl and not at all for disulfide. KAWAMURA (1960) found that the changes in distribution of protein-bound SH during the first cleavage in *Arbacia punctulata* (Fig. 16) closely paralleled those seen previously in *Hemicentrotus pulcherimus, Pseudocentrotus depressus, Mespilia globulus,* and *Clypeaster japonicus* (KAWAMURA and DAN 1958). In addition, the distribution of S-S groups was investigated. Of great interest was the fact that the astral centers, spindle and centrosphere stained no more strongly for SH groups after treatment with reducing agent than the same regions of untreated mitotic figures. However, the egg cytoplasm of all stages almost to metaphase stained uniformly more deeply for SH groups in material in which the S-S bonds were reduced than in material not treated with reducing agent.

2. Isolated mitotic apparatus

A family of methods for isolating the structurally intact mitotic apparatus was developed (MAZIA and DAN 1952, MAZIA 1955, MAZIA, MITCHISON, MEDINA and HARRIS 1961, KANE 1962, MIKI 1962). This opened avenues to the study of the mitotic apparatus by basically different techniques. Information about crosslinks holding the structure together can be obtained from: 1. the nature of the condition efficacious for its isolation and 2. chemical analysis of the isolated division figures themselves. Successful isolation of the mitotic apparatus was contigent upon its selective stabilization under conditions that would allow the subsequent dispersal of the enveloping cytoplasm. In earlier methods cold 30% ethanol was used to stabilize the mitotic apparatus and digitonin to disperse the cell surface and cytoplasm, liberating the morphologically "normal" mitotic figure. It soon became evident that the stability of the mitotic apparatus isolated by the alcohol-digitonin method was very variable from one preparation to the next. Earlier solubility studies on isolated mitotic apparatus (MAZIA and DAN 1952, MAZIA 1955) suggested that it was an S-S bonded structure. This view received further support from in vivo studies on blockage and disorientation of the mitotic apparatus by mercaptoethanol (MAZIA and ZIMMERMAN 1958).

ZIMMERMAN (1960) has made an extensive physicochemical analysis of mitotic apparatus isolated from *Strongylocentrotus purpuratus* first cleavage metaphase stages by the digitonin method (MAZIA 1955). Using dry weight as a measure of total protein the average protein content per mitotic apparatus was determined to be 0.72×10^{-5} mg. This represents about 12% of the protein in the dividing egg and is identical to the value obtained by MAZIA and ROSLANSKY (1956). Analytical ultracentrifugal analysis of an undialyzed solution of mitotic apparatus dissolved in 0.1 M salyrgan resolved two molecular components. The major, slower sedimenting component had an S_{20} of 3.7 and the minor, faster sedimenting component had an S_{20} of 8.6 (see Fig. 17). After dialysis for 24–36 hours against veronal-HCl buffers at pHs 8.5, 9.0, and 9.5 or phosphate buffers at pHs 7.5 and 8.0, there was one broad peak with an S_{20} of 3.2. The presence of a single peak after dialysis suggests that the solubilized mitotic apparatus material is very labile and can be accounted for by a breakdown or aggregation of molecular components to give a different sedimentation characteristic. Electrophoretic patterns revealed two molecular components possessing different electrical charges (see Fig. 18). The major, slower moving component accounts for over 90% of the total mitotic apparatus material and the minor component less than 10%. Work by MAZIA (1955) has shown that after the minor peak has migrated out of the electrophoretic cell, very little material absorbing at 260 millimicrons remained, suggesting that the RNA found in solutions of dissolved mitotic apparatus is associated with (or is) the minor component. The molecular weight of isolated mitotic apparatus dissolved with salyrgan was determined by the technique involving the "approach to sedimentation equilibrium" and found to be 315,000 ± 20,000.

Improvements in the technique for handling the alcohol-digitonin isolated mitotic apparatus suggested that the integrity of the structure did not necessarily depend upon S-S bonds, although it did not rule out the involve-

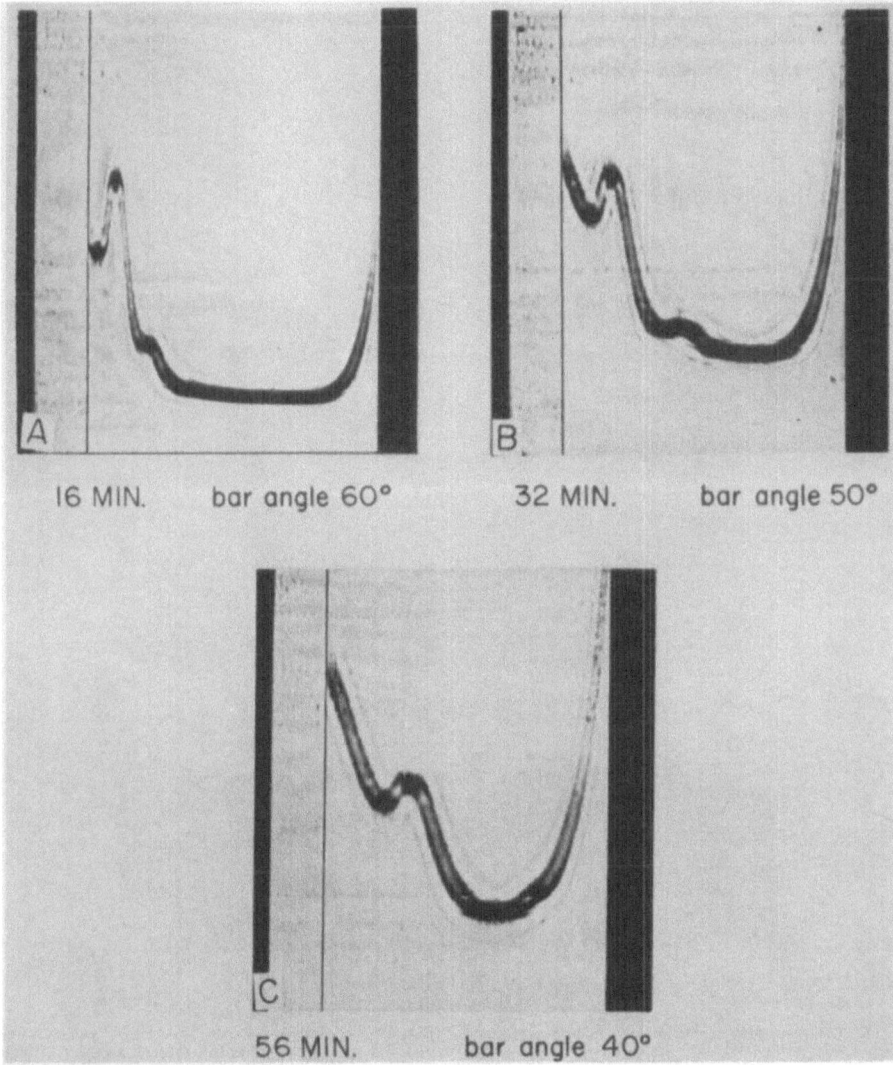

16 MIN.   bar angle 60°   32 MIN.   bar angle 50°

56 MIN.   bar angle 40°

Fig. 17. Ultracentrifuge patterns of mitotic apparatus isolated by the digitonin method from cold ethanol and dissolved in 0.1 M salyrgan (mersalyl acid) at pH 9.0. Centrifuged in Spinco model E analytical ultracentrifuge at 59,780 rpm. A — 16 min., bar angle 60°; B — 32 min., bar angle 50°; C — 56 min., bar angle 40°. The S_{20} for the lighter component is 3.7; for the heavier component it is 8.6. It is evident from C that there is a considerable amount of material that has not really begun to move in the centrifugal field after 56 min. of centrifugation, suggesting that it may have a lower molecular weight than the material responsible for the resolved peaks. This is consistent with later studies (see Fig. 20). From ZIMMERMAN, Exp. Cell Res. 20, 1961, 529—547.

ment of protein bound SH groups. Consistent with this viewpoint are: 1. the conditions under which the mitotic apparatus can be isolated directly from living eggs and 2. the observations on the stability and solubility of the division figures isolated directly from living material (MAZIA,

MITCHISON, MEDINA and HARRIS 1961). An essential ingredient of the isolation medium used to isolate mitotic apparatus directly from living cells is dithioldiglycol, an oxidized form of mercaptoethanol. When this was absent from the isolation medium, intact mitotic apparatus could not be recovered.

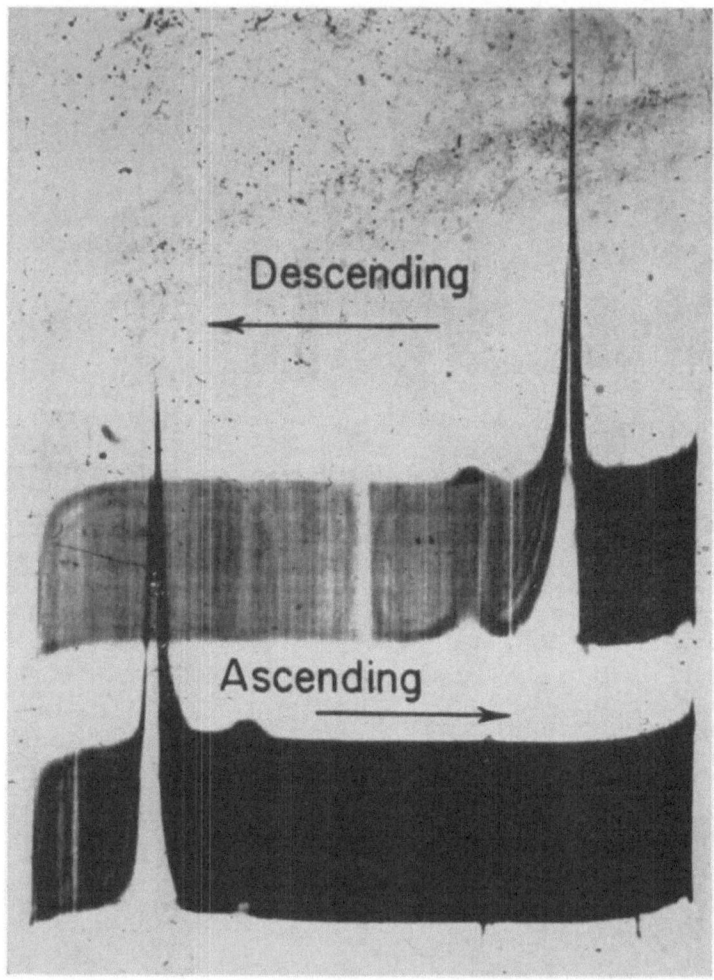

Fig. 18. Electrophoretic patterns of mitotic apparatus isolated by the ethanol-digitonin method and dissolved on 0.1 M salyrgan at pH 9.0. The electrophoresis took place at pH 7.5 in phosphate buffer with an ionic strength of 0.1. The upper pattern is the descending and the bottom pattern is the ascending boundary. From ZIMMERMAN, Exp. Cell Res. 20, 1960, 529—547.

It is thought that dithioldiglycol stabilizes the structure by reversibly increasing the number of S-S bonds. The hypothesis on which this isolation technique is based imagines that a major portion of the intermolecular bonding responsible for maintaining the structural integrity of the mitotic apparatus involves linkages other than S-S bonds between the thiol groups of adjacent molecules. These could be hydrogen bonds or cationic bridges through SH groups. The weakness of these bonds was felt to present no obstacle to their being candidates for this role, if the total number was great enough. This interpretation is also consistent with the observations on the ease with which the mitotic apparatus (isolated directly from living material) can be oxidized to an extremely stable state that can be reversed only with strong reducing agents under alkaline conditions. Pairs of SH groups in close proximity can account for this behavior. MAZIA et al. (1961) suppose that i n v i v o there is a probability that a certain proportion of the pairs of SH groups will be fully oxidized to the S-S form. These are imagined to be in a constant flux between the reduced and oxidized state, so that the over-all stability of the mitotic apparatus reflects the number of SH pairs that are linked at any given moment. They propose that a certain number of SH pairs may even be half oxidized; and, as a consequence, imagine the mitotic apparatus as a giant free radical with constant virtual flow of electrons within it, and also between it and the surrounding systems capable of participating in the flow.

Another interesting feature of the mitotic apparatus isolated directly from living material and observed by MAZIA et al. (1961), is that it can be irreversibly stabilized by magnesium ions and calcium ions in low concentrations.

Mitotic apparatus have also been successfully isolated directly from living cells in a medium of 1 M hexanediol at pH values ranging from 5.5 to 7.0, with values of 6.0 to 6.4 yielding the most normal-looking figures (KANE 1962). The solubility properties of the figures isolated by this method appear to be similar to those for the mitotic apparatus isolated directly from living cells in a medium containing dithioldiglycol. KANE suggests that hexanediol and dithioldiglycol may be functionally equivalent and that the effectiveness of dithioldiglycol is not due to the disulfide group but is due to the speed with which it penetrates the cell and its subsequent effect on cell lysis.

At this point the reader's attention is directed to an important difference between the dithioldiglycol technique (MAZIA, MITCHISON, MEDINA and HARRIS 1961) and the hexanediol technique (KANE 1962) for isolating the mitotic apparatus directly from living sea urchin eggs. More specifically, it is the composition of the medium in which the isolation is done that can be of great importance in determining the chemical character of the isolated mitotic apparatus. Thus, any subsequent biochemical analyses of the isolated mitotic apparatus must be interpreted in relation to the isolation procedure itself. First we'll consider the character of the dithioldiglycol isolation medium developed by MAZIA et al. (1961). It consists of 1.0—1.15 M sucrose, 0.15 M dithioldiglycol and 0.001 M versene, carefully maintained between pH 6.0–6.2. In reference to the dissolved solute concentration, this medium is somewhat hypertonic to the sea urchin egg. The principle osmotically active solute species is sucrose which is a nonpenetrating solute. The dithioldiglycol can be expected to penetrate the cell fairly rapidly. However, since the bulk of the osmotically active solute particles of the medium are nonpenetrating, the cells and any osmotically active cytoplasmic particles (e. g. yolk particles) may achieve equilibrium with very little change in their original volume. Thus, it can be expected that any yolk granules and other osmotically active particles that may be occluded in the isolated mitotic apparatus will be intact. Also the extraction of any lipid-like constituents from the mitotic apparatus may be expected to be kept to a minimum. On the other hand the nature of the hexanediol medium for the isolation of mitotic apparatus developed by KANE (1962) is fundamentally different. It is a 1 M hexanediol solution adjusted to pH 6.0–6.3. Since hexanediol is a rapidly penetrating solute and the isolation medium contains no nonpenetrating solutes, the cells and any osmotically active cytoplasmic particles will undergo osmotic lysis. One might also expect a somewhat greater degree of extraction of lipid-like material from the mitotic apparatus than can be expected to occur during isolation by the dithioldiglycol method (which should already be very little). In view of these considerations, biochemical studies on isolated mitotic apparatus should be performed on figures isolated by both techniques.

Biochemical studies on mitotic apparatus isolated by the ethanol-digitonin method can be expected to yield a very warped picture of the lipid components that may be important molecular constituents of this structure.

From the foregoing it seems clear that structural proteins bearing SH groups are abundant in the mitotic apparatus, but that there is no definite concept of how these proteins are cross-linked to form the mitotic apparatus. Good cytological evidence has revealed fluctuations in the protein SH level of the region occupied by the mitotic apparatus during the course of division. Changes in S-S bonding are also reported. It is suspected that many weak bonds can account for the native stability of this structure and that pairs of SH groups proximal to each other can be oxidized to S-S bonds, imparting to the isolated mitotic apparatus the greatly increased stability so frequently observed.

It is tempting to speculate that the combined efforts of many weak bonds (i.e. hydrogen bonds) are responsible for the overall rigidity and the visible structural organization of the in vivo mitotic apparatus at any given moment; and further that the smooth transition through successive structural transformations associated with mitosis reflect the localized sequential making and breaking of different, but stronger, bonds considerably fewer in number. These can be imagined to be sulfhydryl-disulfide interchanges between adjacent protein molecules. Such an interpretation is consistent with the observation that isolated, structurally "native" mitotic apparatus have not been seen to move chromosomes. The very groups required to impart functional activity to this structure under normal conditions (the protein bound SH) are diverted to impart the necessary structural rigidity to the isolated mitotic figure. This implies that the visibly unstructured region of the in vivo mitotic apparatus provides the structured region with a general environment which retains thiol groups in the reduced state, but that local changes occur during mitosis which allow these to become reversibly oxidized to disulfide bonds. The structured region of the mitotic apparatus contains the mechanism by which the physical movements are effected; and the enveloping unstructured area provides the means for setting this mechanism into motion.

3. Effect of various chemical treatments on the organization of the mitotic apparatus

There is a diverse array of chemical treatments which can induce structural modifications of spindles and asters i n v i v o. Many of these appear to disorganize the mitotic apparatus by disrupting the intermolecular crosslinks holding the molecular subunits to each other, rather than by changing the chemical configuration of the subunits. The effects of colchicine on the mitotic apparatus and on chromosome behavior have been extensively studied and are reviewed by EIGSTI and DUSTIN (1955). Reference here will be made only to the work of GAULDEN and CARLSON (1951), who observed that colchicine disrupted a completely formed spindle in *Chortophaga viridifasciata* and suggested that the spindle material itself was not destroyed, but merely modified in its molecular orientation. The effect of colcemide, which is structurally closely related to colchicine, applied to *Strongylocentrotus purpuratus* eggs at metaphase of the first cleavage division has been investigated by SAUAIA and MAZIA (1961). Their observations on mitotic apparatus, isolated by the alcohol-digitonin method from eggs removed after varying exposure times to colcemide, revealed that different regions of the mitotic apparatus were not simultaneously disorganized to the same extent. After a four minute exposure to 0.001% colcemide the astral rays were reduced or absent, but the spindle fibers were still visible. A seven minute exposure resulted in the loss of the centers, but some spindle fibers were still evident. After ten minutes, the spindles had lost nearly all of their fibrous organization and were reduced in length. Small spherules, associated with chromosomes, and spherical bodies were isolated from eggs exposed for longer periods to colcemide. The significance of the spherical bodies is not known. Similar phenomena have been seeen in other experiments.

Dithioldiglycol was added to the sea water of developing sand dollar eggs (*Dendraster excentricus*) at various times from prophase to late metaphase. The resultant structural disorganization of the mitotic apparatus followed a pattern which was at least superficially similar to that described for the shorter colcemide treatments (WENT 1962). The asters were the first region to be effected. Their reduction in extent was closely followed by the loss of their fibrous organization and occurred before the spindles themselves had completely lost their fibrous appearance. Observations on living material treated with dithioldiglycol clearly revealed a radially fibrous organization of the asters, although less pronounced than that seen in the control eggs. In the same experimental material preserved in Carnoy's these regions were devoid of radially fibrous organization. No such discrepancy between preserved and living material has been observed in control cells. Thus the radially fibrous organization of the asters seen in living cells treated with dithioldiglycol was not observed after the material had been preserved in Carnoy's. It can be inferred that the dithioldiglycol interferred with the stabilization of the radially oriented substructures of the asters seen in living cells, while allowing these substructures to assume a limited radial configuration. The apparent discrepancy between the s t a b i l i z i n g influence of relatively high concentrations of dithioldiglycol (0.15 M) employed in the medium used to isolate the mitotic apparatus directly from living material (see Mazia et al, 1961) and the d i s o r g a-n i z i n g influence of lower concentrations (about 5×10^{-4} M) just described has not been resolved.

I n v i v o treatment of developing loach eggs with ATP and sodium sulfate was observed to cause the rapid loss of the fibrous organization of the mitotic apparatus (Poglazov 1961).

Features shared by all of the previously mentioned studies on mitotic aberrations induced by chemical means is that mitosis would continue to a certain point and abruptly stop; or that mitosis would proceed more slowly than normal to completion, provided that the exposure was not excessive. Deviation from this more usual pattern was observed by Gross and Spindel (1960 a, b) when they inhibited cell division in *Arbacia punctulata* by heavy water (D_2O) as an approach to elucidating the mechanism of division. They made the interesting observation that cells in mitosis immersed in D_2O-sea water remained cytologically unchanged, regardless of the stage in division at the time of immersion. Cytokinesis was also arrested. Thus, D_2O appeared to "freeze" the spindles, asters and cleavage furrows in the state extant at the time of immersion. They think that he stabilization of the achromatic figure and of the cleavage furrow was very likely related to the fact that relatively low D_2O concentration caused a sharp increase in cytoplasmic viscosity (measured as stratification time) in non-dividing eggs. A working hypothesis based upon a consideration of their observations, in conjunction with known physical and chemical properties of D_2O, was proposed. This viewed the antimitotic effect of D_2O as stemming from an increased rigidity of macromolecular superlattices in the cytoplasm including the immobilization of the mitotic apparatus. They point out that one consequence of this hypothesis is to ascribe great importance to the role which hydrogen bonds play in maintaining the orderly structure of the achromatic figure. Marsland and Zimmerman (1963) have confirmed the observations of Gross and Spindel (1960 a, b) with respect to the immediate arrest of karyokinesis by D_2O. However, they point out that the cell cortex retains the capacity to perform mechanical work in cleaving the egg.

From some very recent studies on the suppression of mitosis by D_2O in onion roots (Bal and Gross 1964) there emerges a picture somewhat different from that observed for sea urchins. Two differences can be noted: 1. the time required for the onset of mitotic inhibition is much longer in the root cells than in sea urchin eggs, 2. in root tip cells, the mitoses in progress at the time of exposure to the heavy water are completed before the cells were arrested in the following interphase.

4. Other molecular types associated with the mitotic apparatus

The mitotic figure contains other molecular species in addition to the proteins with sulfhydryl groups. In cytochemical studies on *Paracentrotus lividus,* Immers (1957) found material in the mitotic apparatus which stained intensely by means of the periodic acid-Schiff (PAS) technique. The spindle poles stained more strongly with PAS at the transition from prophase to metaphase than earlier in prophase, indicating changes in carbohydrate components at the spindle poles. The staining intensity gradually decreased and disappeared by telophase. The presence of polysaccharides in the mitotic spindle detected by cytochemical methods is also reported by SHIMAMURA and ÔTA (1956), who noted considerable variation in degree of staining among different species.

Nucleic acids, other than the DNA of the chromosomes, have been detected in the mitotic apparatus by cytochemical methods. In studies on mitosis in cultures of newt tissues, Boss (1955) reports that the ribonucleo-protein concentration appeard to be highest in the equatorial region between the separating daughter chromosomes at anaphase. SHIMAMURA and ÔTA (1956) present a table summarizing published reports on the presence of pentosenucleic acid (PNA) in the mitotic spindle. This table reveals that mitotic apparatus containing PNA have been found in: root tip cells of *Vicia, Allium* and *Tradescantia;* pollen mother cells in *Tradescantia, Allium, Lilium,* and *Vicia;* eggs of amphibians, sea urchins, *Ascaris* and *Cyclops,* grasshopper spermatocytes; and cultured fibroblasts of chick embryos. In their own observations on the distribution of PNA in the mitotic apparatus in cells of various plant species, SHIMAMURA and ÔTA noted that it accumulated in polar caps at late prophase. The metaphase spindle had abundant PNA and it was clearly distinguishable from the cytoplasmic area. They interpret some of their observations to mean that the interzonal region of the mitotic apparatus differs considerably in compo-sition from the poleward regions. The phragmoplast was rich in PNA. RUSTAD (1959 d) employed a somewhat different approach to demonstrate the presence of RNA or an RNA-like material in the mitotic apparatus. He stained mitotic apparatus, isolated by the alcohol-digitonin method, with gallocyaninchromalum and Azure B, and observed that the polar regions of the asters at metaphase and early anaphase stained more intensely than any other region of this structure. In middle anaphase the interzonal region displayed a low stainability, but by late anaphase this region stained much more intensely. RNAase treatment removed almost all of the stainable material from the mitotic apparatus, except that present in the chromo-somes.

With cytophotometric measurements on material stained with gallo-cyanine-chromalum molecules, STICH and McINTYRE (1958) estimated the RNA content during development in *Cyclops strenuus* eggs, especially in regard to spindle formation. RNA accumulated in the nuclear sap during the period before the onset of prophase. Following this, the protein content of the nuclear sap increased. In view of the importance of RNA in cellular protein synthesis, they suggest that the accumulation of RNA in the nuclear sap was the cause of the subsequent increase in protein content. This correlation implicates the presence of a third protein-synthesizing system in the nuclear sap, in addition to the nucleoli and the chromosomes. In summarizing their data and that published earlier (STICH 1951, 1954 a, 1954 b), they show that spindle formation was preceded by the accumulation of polysaccharides in the nuclear sap during late interphase, followed by the appearance of RNA and the subsequent increase in protein concentration. Spindle fibers could appear only after these synthetic processes were completed and they entertain the possibility that purine anti-metabolites and U. V. radiation induce mitotic inhibition by attacking the RNA-protein synthesizing system which must function for the formation of the mitotic apparatus. Metabolic inhibitors were seen to interfere with specific phases of the events leading to meiosis (STICH 1954 b). 2,4 dinitrophenol and sodium azide, which both inhibit the formation of energy rich phosphates, prevented the accumulation of polysaccharides and RNA in the nucleus and the subsequent formation of the meiotic spindle. On the other hand, potassium cyanide and monoiodoacetic acid were without effect on polysaccharide and RNA metabolism of the nucleus, but cleavage was completely suppressed. It was concluded that the division of the nucleus depended upon energy rich phosphate and that respiration was important to cell division only in so far as it provided the necessary energy rich phosphates. STICH (1954 b) then presents a scheme to account for the changes in the nuclear sap from prophase until metaphase. The first event is the appearance of rigid RNA molecules, along which protein molecules become aligned to form RNA-protein complexes. These then become aligned end-to-end to form a labile radial orientation initiated in the proximity of the spindle pole. This labile array is then stabilized by polysaccharide molecules.

The isolated mitotic apparatus gives a positive PAS reaction (MAZIA 1955) but the polysaccharide has not yet been isolated by chemical means.

It should be possible to follow the changes in RNA distribution in the mitotic apparatus in living cells by ultraviolet microscopy. The technical problem of preventing damage to cells in mitosis which are under continuous ultraviolet illumination was surmounted by MONTGOMERY and BONNER (1959) who used the flying spot principle in conjunction with television viewing. Their photographs do not reveal conspicuous accumulation of ultraviolet absorbing material in the interzonal region at anaphase or telophase in newt endothelial cells in culture. The apparent absence of RNA from the interzonal region of living cells observed with the flying-spot microscope could result from the presence of other ultraviolet absorbing material that masks the existence of RNA, and need not conflict with the reported presence of RNA in the interzonal area of other cell types.

The analysis of isolated mitotic apparatus consistently reveals RNA associated with this structure. ZIMMERMAN (1960) has analyzed the mitotic apparatus of sea urchin eggs isolated by the alcohol-digitonin method and found that it contained about 6% RNA. The molar ratio of the ribonucleotides were not consistently different from those observed for the RNA of the whole cell. MAZIA (1961) points out that, although we associate RNA with protein synthesis, there are other possible roles for the RNA associated with the mitotic apparatus. One idea is that the RNA serves as an energy-storing device which implies that the assembled mitotic apparatus is activated and can perform its act with no further intervention from exogenous energy supplies. However, this attractive suggestion must be modified in view of more recent experimental data. EPEL (1963) studied the effects of carbon monoxide inhibition on ATP level and rate of mitosis in sea urchin eggs. From this he derived the following generalizations: (1) mitosis in the cleaving sea urchin egg can be blocked at a n y stage if inhibition is applied at an appropriate time before that stage. Therefore any reserves on hand at the initiation of inhibition permit division to proceed for only as long as they last at which time division is arrested, (2) the amount of ATP in the sea urchin egg does not change during the cell cycle, (3) mitosis stops when the total ATP level drops to 50% of normal. AMOORE (1963) has shown that mitosis can be arrested while it is in progress in pea root tips by respiratory inhibitors.

Another very tenuous idea assigns to RNA the role of mutual recognition required of the molecular subunits in order that they can become assembled into the definitive mitotic apparatus.

It is clear that the mitotic apparatus contains RNA, presumably complexed with protein, but we do not know the significance of its presence.

DIRKSEN (1961 a) reports that there may be as much as 40% lipoprotein in the mitotic apparatus isolated directly from living material. This is not found in alcohol-digitonin isolated mitotic apparatus. The presence of lipoprotein may account for the observation by SWANN (1954) that the fibrous organization of the mitotic apparatus can be abolished by ether.

KAWAMURA and DAN (1958) observed that treatment of *Hemicentrotus pulcherimus* with sea water containing more than 4% ether resulted in the suppression of both asters and spindle and failure of both karyokinesis and cytokinesis. On the other hand eggs subjected to 0.6% ether-sea water underwent karyokinesis but not cytokinesis owing, probably, to the failure of aster formation.

Specific tests for masked lipids has shown that the centromere and the outer pellicle of the spindle tubules in anaphase stages of *Allium cepa* and *Vicia faba* root tip cells, *Agapanthus umbellatus* pollen mother cells and *Cavia cobaia* first order spermatocytes has revealed the presence of lipids in these structure (SERRA and SEXIAS, 1962). SERRA and SEXIAS suggest that their observations favor the view that the similarity in composition between the spindle fibers and the centromere is the basis for the attachment of the fibers to the chromosomes.

5. Effect of metabolic inhibitors; enzyme activity

The effects of specific inhibitors of protein synthesis on the dynamics of spindle formation in living new tissue culture cells have been investigated by TAYLOR (1959). In this material the spindle forms between the asters while the latter are separating. The spindle length increases from 8μ to 34μ at a uniform rate of about 1.43μ per minute over a period of 15 to 18 minutes. It was reasoned that in growing cells at least half of the protein for the spindle must be synthesized for each division. Chloramphenicol and two amino acid analogues, p-fluorophenyl alanine and beta-2-thienyl alanine, were tested. The amino acid analogues did not prevent division and spindles of normal length were formed at the usual rate for at least two or three hours after the inhibitors were added. However, the effects of chloramphenicol were striking. Cells which had begun to form a spindle completed division at a normal rate, while cells which had not initiated spindle formation produced shorter than normal spindles or none at all. The longer the exposure to chloramphenicol prior to the appearance of the asters, the shorter the spindle. Normal anaphase was observed even when the spindle length had been reduced to about one-half of the normal length. The chloramphenicol effect was interpreted to mean that spindle protein synthesis occurred during the one hour period preceding spindle formation and was completed before the spindle became visible. However, when the synthesis of the spindle protein was blocked, the cell constructed the spindle only from the protein already present before the drug was added. This would account for decreasing spindle size with increasing exposure to chloramphenicol. However, in a subsequent report TAYLOR (1963), presents data obtained from mammalian culture cells, that require a different interpretation. It was shown that a fraction of the cells continued to enter metaphase and anaphase with normal spindles for several hours after the addition of chloramphenicol. TAYLOR suggests that the chloramphenicol has apparently two effects on mitosis: (1) a direct effect on the mitotic apparatus of those cells which enter mitosis s h o r t l y after exposure to cloramphenicol, and (2) an inhibition of the synthesis of the proteins necessary for the mitotic apparatus in those cells exposed to the chloramphenicol for a l o n g e r period before mitosis.

By autoradiographic studies on isolated mitotic apparatus, STAFFORD and IVERSON (1963) demonstrated that leucine-C[14] added after fertilization was incorporated into some protein of the mitotic apparatus of *Lytechinus variegatus* and *Echinometra lucunter* (Fig. 19). That the observed activity was actually from label incorporated into protein of the mitotic apparatus was demonstrated by the following control experiments: (1) the cells were exposed to leucine-C[14] for fifteen minutes beginning five minutes after fertilization, after which they were washed in excess leucine-C[14] and resuspended in sea water containing excess leucine-C[14] until the mitotic apparatus were isolated by KANE's technique (1962). The isolated mitotic apparatus were washed extensively (after stabilization) with excess leucine-C[14] before autoradiography; (2) mitotic apparatus isolated from cells grown continuously in leucine-C[14] were washed and then treated with trichloroacetic acid at 90⁰ C for fifteen minutes to remove nucleic acids and allow any protein associated with the transfer-ribonucleic acid of ribosomes "caught" in the isolated mitotic apparatus to be washed away. GROSS and COUSINEAU (1963) have also demonstrated the incorporation of DL-leucine-H[3] into early prespindle, prophase and metaphase regions in *Arbacia punctulata*. Among the valid reasons they offer for believing that the label is intimately associated with the mitotic apparatus is that the activity becomes generally distributed only after the mitotic apparatus has broken down, and the subsequent return of activity to the mitotic figures. They suggest that synthesis of small amounts of protein in the cytoplasmic matrix could produce the functional fibrous component of the mitotic apparatus which enmeshes a large quantity of ground cytoplasm. In very recent work, Mangan et al. (1965) localized the site of H^3-leucine incorporation into the mitotic apparatus of *Arbacia punctulata*. They

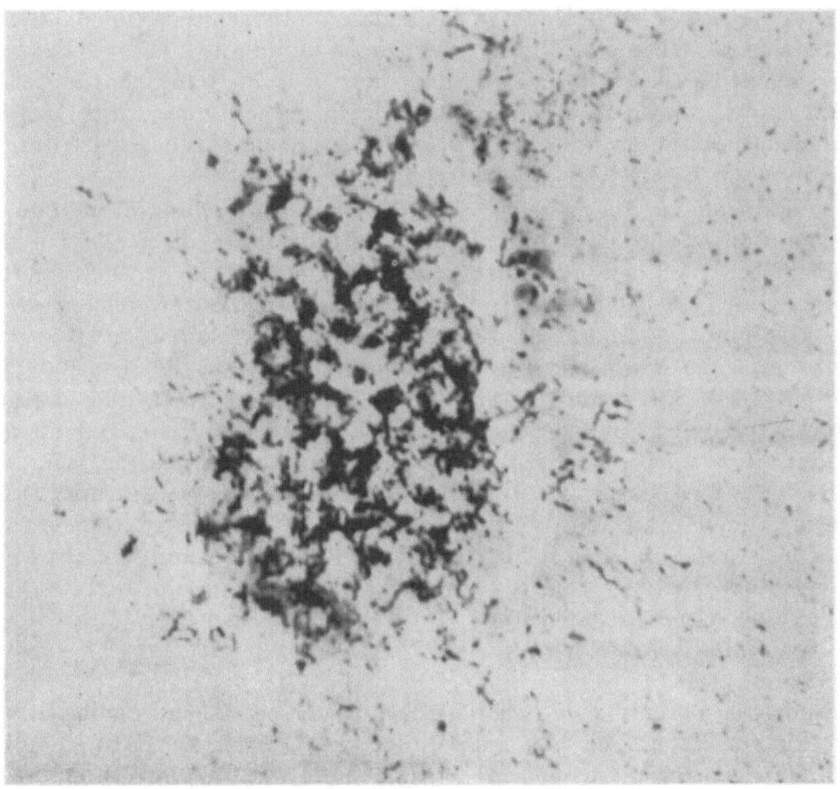

Fig. 19. Radioautograph of mitotic apparatus isolated by Kane's (1962) technique from *Echinometra lucunter* egg at first cleavage. The cell had been exposed to sea water containing 10⁻⁷ M leucine-C14 (10.2 mc/mM) from after fertilization until metaphase was visible. The isolated mitotic apparatus had been treated with hot trichloracetic acid. Radioautograph exposed for 7 days. From Stafford and Iverson, Science 143, 1964, 580—581.

observed in their electron microscope autoradiograms a close, non-random association between the silver grains and the tubular spindle filaments.

The incorporation of labelled amino acids into cytasters in sea urchin eggs has been studied by autoradiographic techniques. Gross, Spindel and Cousineau (1963) report that about one hour after unfertilized *Arbacia punctulata* eggs were placed in 85% D_2O the cytoplasm was packed with cytasters. When these eggs were washed free of D_2O medium, they became activated. That is, the fertilization membranes elevated and the cells divided. Cleavage began in about one-half of the normal time and continued for about eight hours. Although the furrows segmented the eggs randomly, mitoses were mostly normal. These "embryos" then disintegrated without visible differentation. The cytasters induced by exposure to D_2O containing medium became heavily labelled, as judged by autoradiography, when they were allowed to form in the presence of labelled amino acids. When they measured protein synthesis during this same period it was found that it was two to five times greater than that of normal unfertilized eggs in normal sea water and the peak of synthesis occurred just before the full development of the cytasters.

In experiments with HeLa cells ERRERA and BRUNFANT (1964) observed the pattern of tritiated cytidine incorporation into the mitotic figures of pulse labelled cells. It appeared that there was no significant uptake of cytidine into spindle RNA during spindle formation from which fact they concluded that the spindle RNA appears all to be synthesized before the appearance of the mitotic apparatus.

In addition to the already mentioned recent observations on the incorporation amino acids into the mitotic apparatus, there are a number of other recent reports indicating an abrupt increase in synthetic activity of sea urchin eggs immediately following fertilization. Among these are the activation of amino acids during development (MAGGIO and CATALANO, 1963); the formation of polyribosomes upon fertilization (MONROY and TYLER, 1963; STAFFORD, SOFER, and IVERSON, 1964); the early acceleration of protein synthesis following fertilization which does not appear to depend upon the synthesis of new messenger RNA (GROSS, MALKIN and MOYER, 1963); and that the net in vivo formation of polyribosomes requires a continual supply of energy under conditions where messenger-RNA synthesis has been shown not to be a rate limiting factor (HULTIN, 1964). These and the previously discussed observations clearly demonstrate the relatively high levels of synthetic activities initiated upon fertilization in sea urchin eggs, although there is normally no net requirement for an exogenous source of structural precursors.

Biological motility can include chromosome movement during mitosis. In seeking unity among biological processes it would seem reasonable to find ATPase activity associated with the mitotic apparatus. With this in mind, MAZIA, CHAFFEE and INVERSON (1961) examined isolated mitotic apparatus for ATPase activity. Those isolated by the alcohol-digitonin method were inert in this respect, but mitotic apparatus isolated directly from living material contained an active and characteristic ATPase. This enzyme was highly specific for ATP; guanosine triphosphate, cytidine triphosphate, and uridine triphosphate were not split at all. No pyrophosphatase, alkaline phosphatase or ADP-splitting enzymes were found. They could not show unequivocally, however, that the ATPase was associated with the molecules responsible for the "fibrous" structure of the mitotic apparatus. ATPase activity has also been reported in the mitotic apparatus of other sea urchin species. MIKI (1963) fractionated anaphase stages of *Anthocidaris crassispina* into the mitotic apparatus fraction and the cytoplasmic fraction and measured the inorganic phosphate liberated by these two fractions. When the results were expressed as $\mu g \, P/mg \, N/15$ min., the ATPase activity of the mitotic apparatus fraction was three times greater than that found in the cytoplasmic fraction. By histochemical methods, it was shown that metaphase and anaphase mitotic apparatus had higher ATPase activity than the enveloping cytoplasm. It was suggested that part of the SH groups seen by KAWAMURA and DAN (1958) belong to the ATPase.

MIKI (1964) localized ATPase activity in sea urchin (*Anthocidaris crassispina)* eggs from fertilization to the two cell stage by cytochemical techniques. The unfertilized egg showed a uniform staining throughout the cytoplasm. However, at the monaster stage, the monaster and the astral rays are deeply stained. At the streak stage, the nucleus and the "streak" are darkly stained and the cytoplasm surrounding the "streak" is more darkly stained than the peripheral cytoplasm. The metaphase mitotic apparatus was very deeply stained. By anaphase, the asters were still clearly more strongly stained than the surrounding cytoplasm, but the spindle was not visible. The staining pattern of the resting stage following cleavage closely resembled that seen five minutes after fertilization.

C. Immunochemical evidence

This approach became feasible only after techniques for the mass isolation of the mitotic apparatus from sea urchin material were developed. The initial immunochemical studies were designed to demonstrate the presence or absence of antigens in the unfertilizied egg which are immunochemically identical to antigenes found in solutions of dissolved mitotic apparatus (WENT, 1959). The results revealed that the unfertilized egg contains an antigen immunochemically identical to an antigen (termed the precursor-1 component) which is invariably present in all solutions of dissolved isolated mitotic apparatus investigated, irrespective of the isolation method used. However, with the advent of the direct isolation method (MAZIA, MITCHISON, MEDINA and HARRIS, 1961) and the consequently milder conditions under which the isolated figures could be dissolved, solutions were obtained that contained another immunochemically identifiable antigen shared with unfertilized eggs. Later evidence obtained with antiserum of higher activity (data of SAUAIA; see MAZIA, 1961: p. 141), indicated the presence of at least four antigens in solutions of dissolved mitotic apparatus. Of course, it is not known how many of these antigens are actually directly reponsible for the fibers seen in the isolated figures. It is entirely possible that the "fibers" of the mitotic apparatus are composed of only one antigen species and that the other antigens present in the solutions of dissolved mitotic apparatus are in the visually unstructured regions of the mitotic apparatus.

This is not to imply that they are unimportant in the overall functioning of the division figure. The immunochemical evidence is entirely consistent with the view expressed by INOUÉ (1960) that the visibly structured areas of the mitotic apparatus are bathed by a pool containing a random array of the same molecules.

Immunology has also shown that the asters and the spindle of the mitotic apparatus appear to have the same antigenic composition. Using the Ouchterlony technique SAUAIA and MAZIA (1961) compared the antigenic composition of spindles isolated from material treated with colcemide with that of whole mitotic apparatus and obtained band patterns which indicated that the mitotic apparatus contained no antigens that were not present in the spindle. The very reasonable hypothesis that cytasters were antigenically indistinguishable from the mitotic apparatus was investigated by Dirksen (1961 a, 1964). She prepared solutions of cytasters induced by Loeb's "double method" and isolated by the alcohol-digitonin technique. When these were compared immunochemically to solutions of dissolved mitotic apparatus, no differences in antigen composition were noted.

When it had been demonstrated that the only antigens (including the precursor-1 component) detectable in solutions of dissolved mitotic apparatus were also present in the unfertilized eggs, a study of the intracellular distribution of the precursor-1 component in the unfertilized egg was undertaken. It was thought that this might provide insight into the sequence of events leading to the mobilization of the precursor-1 antigen into the mitotic apparatus. The p a r t i c u l a t e   f r a c t i o n was obtained from unfertilized eggs and extracted by various means. (The particulate fraction was arbitrarily defined as everything that can be sedimented from a homogenate of unfertilized ethanol-preserved eggs at 60,000 g for 30 minutes.) This was then used to absorb suitable antisera. The results demonstrated that there was an antigen tenaciously associated with the particulate fraction which could combine specifically with the antibodies homologous for the precursor-1 component. Then careful fractionation of living unfertilized eggs in isotonic dextrose was undertaken in the attempt to resolve the intracellular distribution of the precursor-1 component (Went, 1960). None of the fractions studied lacked the precursor-1 component. These included: (1) the contents of osmotically lysed yolk particles; (2) 0.5 M KCl extract of the residual solid matter remaining after the yolk particles were lysed; (3) 0.5 M KCl extract of the mitochondria-microsome fraction. Nuclei were not investigated. We thus find the precursor-1 component to be widely distributed throughout the egg; both in an easily extractable form and in a form tenaciously associated with cytoplasmic structures.

It is important to point out here that it is not known what fraction the precursor-1 component is of the total protein in a solution of dissolved mitotic apparatus, but initial semi-quantitative estimates (unpublished) suggest that no more than 50%, and possibly less, of the total protein is the precursor-1 component. The exact relationship of this protein to the mitotic apparatus is also not known.

Experiments (Went, 1959) designed to detect an immunochemical relationship between the precursor-1 component and soluble proteins extracted from adult sea urchin muscle were negative. In support of this, Holtzer et al. (1959) reported that fluorescein isocyanate-labelled antibodies against chick myosin did not react with the mitotic apparatus of chick fibroblast cells.

The immunochemical data allow us to formulate the following tentative viewpoints of the sea urchin mitotic apparatus: (1) the formation of the mitotic apparatus can, in part, be viewed in terms of the spatial rearrangement of pre-existing molecules in the unfertilized egg; (2) cytasters, spindles, and asters are immunologically indistinguishable from the whole mitotic apparatus; (3) the precursor-1 component, implicated in the visibly structured regions of the mitotic apparatus, is ubiquitous in its intracellular distribution; (4) in no way do these data conflict with INOUE's opinion that the fibrous regions of the mitotic apparatus may be areas of relatively high micellar organization bathed by a pool of the same, but randomly oriented, micelles.

There follows from this the question of whether or not the mitotic apparatus contains a unique molecular species not present in the cytoplas-

Fig. 20. Disc electrophoretic patterns of dissolved mitotic apparatus (gels A and C) and extracts of cytoplasmic particles (gels B and D). Gels A and B show the band pattern in 5% gel columns; gels C and D the pattern in 7½% gel. The material responsible for the entire pattern seen in the 5% gels is restricted to the very dense, undifferentiated band at the very top of the respective 7½% gels. Thus, none of the material responsible for the bands below the dense undifferentiated band at the top of gels C and D remains in gels A and B; having migrated completely through the gels into the buffer solution. The material designated the precursor-1 (WENT, 1959) does not form a discrete band in acrylamide gel, instead it diffusely occupies a region in the top 4 mm. or so of the 5% gel-this would extend just below the second dense band from the top in gel A. From these band patterns it can be seen that there is no band in the electrophoretic pattern of dissolved mitotic apparatus that cannot also be seen in the corresponding pattern from the extract of cytoplasmic particles.

mic region of the cell enveloping the mitotic apparatus. Immunochemical data imply that there is no antigen present, at the time of division, which is uniquely restricted to the mitotic apparatus. This is supported by disc electrophoresis in acrylamide gels (WENT, unpublished) of solutions of isolated mitotic apparatus and extracts of the surrounding cytoplasmic particles. The mitotic apparatus were isolated by a slightly modified version of KANE's hexanediol technique KANE (1962), and dissolved under mildly alkaline conditions in low ionic strength. The cytoplasmic particles occupying the remainder of the cell were concentrated and subjected to the same conditions employed for dissolving the isolated mitotic apparatus. The band patterns obtained following the electrophoresis of these solutions in acrylamide gel (Figure 20) indicate that there is no molecular component detectable in the mitotic apparatus solutions which cannot also be detected in extracts of the enveloping cytoplasmic particles. Thus there is currently no evidence for the presence of a unique "mitotic apparatus protein", which, at the time of cell division is restricted in its distribution to the mitotic apparatus. However the search for a mitotic apparatus protein has not been exhaustive and has been restricted to large, water-soluble molecules extractable under mild conditions. It will be very important to investigate carefully the petroleum ether soluble fraction of the mitotic apparatus for unique molecular species owing to the involvement of lipids in biological membranes and the known effects of ether on the i n v i v o mitotic apparatus.

In the attempt to identify the position (WENT, unpublished) of the precursor-1 component (WENT, 1959) in the acrylamide gel it became apparent that this antigen does not form a discrete band, but rather, it diffusely occupies a region in the top 4 mm. of a column of 5% acrylamide gel. Its position was established by eluting thin discs of the gel after electrophoresis and examining the eluates with two dimensional immunodiffusion (OUCHTERLONY technique) in agar sheets. Clearly not all material that can migrate electrophoretically in acrylamide gel will form a discrete band, compounding the difficulty of detecting molecular species unique to the mitotic apparatus. The electrophoretic pattern of dissolved mitotic apparatus in acrylamide gels (Fig. 20) reveals a much greater molecular heterogeneity than can be observed in the ultracentrifugal patterns obtained by ZIMMERMAN (1960) (see Fig. 17). The difference may, in part, be ascribed to the presence of smaller molecular weight molecules that were clearly resolved in the 7½% acrylamide gel but which did not move significantly from the boundary in the ultracentrifuge cell (see Fig. 20). The ultracentrifugal pattern suggests this by the "peak" of unresolved material that remained at or near the surface.

Experiments providing evidence for the activation of ribosomes in sea urchin eggs in response to fertilization (HULTIN, 1961 b), the effect of puromycin on protein metabolism and its inhibition of cell division in *Paracentrotus lividus* (HULTIN, 1961 a) and the previously mentioned experiments of GROSS, et al. (1963) and SAAFFORD and IVERSON (1963) showing the incorporation into the mitotic apparatus of labelled amino acids added after fertilization, all clearly demonstrate the labile nature of at least some of the proteins in unfertilized sea urchin eggs, if not the net synthesis of new protein species immediately following fertilization. It is most likely that the precursor concept (WENT, 1959) cannot entirely account for the constituent molecules of the mitotic apparatus and that some of the molecules essential for the structural and functional integrity of this structure are not present in the unfertilized egg, but must be synthesized during the period between fertilization and the first cleavage. In this connection, MANGAN et al. (1965) observed that the specific activity of proteins, labelled with H^3-leucine, from isolated sea urchin (*Arbacia punctulata*) mitotic apparatus was more than three times that of the proteins from the rest of the cell. Exhaustive purification of these two fractions can exclude from consideration label incorporated into lipids.

D. Evidence from electronmicroscopy

The mitotic apparatus has only stubbornly yielded its secrets to scrutiny with the electron microscope. The early techniques produced images too coarse to add significantly to what was already known from light microscopy. However, the work of recent years has uncovered significant new facts and one may be sure that a great deal more remains yet unseen.

The basic structural components of the spindle, described from observations by light microscopy, i.e. the chromosome-to-pole fibers, the pole-to-pole fibers and the centrioles, have also been seen with the electron microscope (see Figs. 21–25). The latter have elready been discussed elsewhere in this article.

The electronmicrographs of chromosome-to-pole fibers reveal the presence within them of submicroscopic elements, the s p i n d l e   f i l a m e n t s. According to PORTER (1955), spindle fibers have at their center a pair of fine, tubular elements, the spindle filaments, surrounded by a cytoplasmic matrix and dense particles. The spindle filaments were about $25 \, \text{m} \mu$ in diameter. The filaments are also found in the interzonal region and in astral rays. He points out the similarity between these paired filaments in the spindle and the double filament unit of cilia. This is provocative in view of the fact that spindles and cilia are formed under the influence of basically homologous bodies: centrioles and basal bodies respectively. PORTER considers the spindle fiber to consist of a pair of spindle filaments and the immediately surrounding material. This would correspond to the spindle fibers seen with the polarized light microscope. In spindles of dividing mouse spermatogonial cells numerous individual fibers, ranging from $50 \, \text{m} \mu$ to $80 \, \text{m} \mu$ in thickness, are described by SCHULTZ-LARSEN (1953). These fibers were seen to cross, but no branching or anastomosis was mentioned. It was impossible to distinguish between chromosomal, continuous and interzonal fibers. Even the early efforts with electronmicroscopy by BEAMS et al (1950) disclosed that the chromosomal fibers were made up of smaller fibers ranging from $30 \, \text{m} \mu$ to $60 \, \text{m} \mu$ in width.

On the basis of other reports, it appears that spindle fibers seem to conform to the same basic architectural pattern already described by PORTER (1955) and BEAMS et al. (1950). These will be briefly summarized: SATO (1958, 1959) observed that the chromosomal fibers in *Lilium* and *Magnolia* spindles could be composed of as many as 15–20 fine fibrils, apparently tubular, ranging from $20 \, \text{m} \mu$ to $70 \, \text{m} \mu$ in width: NEBEL and COULON (1962) found fine fibrils occurring in loose bundles of 2 to 4 in the spindle region of metaphase I in pigeon spermatocytes; ROTH et al. (1960) and ROTH and DANIELS (1962) described the occurence of spindle fibers, $15 \, \text{m} \mu$ in diameter, in small bundles attached to the chromosomes of *Amoeba proteus;* GROSS et al. (1958) have also detected fibrous, tubular-appearing elements about $20 \, \text{m} \mu$ in diameter which tend to be aggregated into larger bundles; tubular spindle filaments attached to centromeres are described by GATENBY (1959) for spermatogenesis of *Lumbricus* and it is felt that there is no doubt that the spindle filaments in general are tubular.

The spindle region may (ROTH and DANIELS, 1962), Figure 21, or may not (RUTHMANN, 1958) contain ribosome-like particles. However, when these are present, they show no preferential association with the spindle filaments. Large structures, such as mitochondria and yolk granules, are usually excluded from the spindle and astral regions. In contrast, it is not unusual

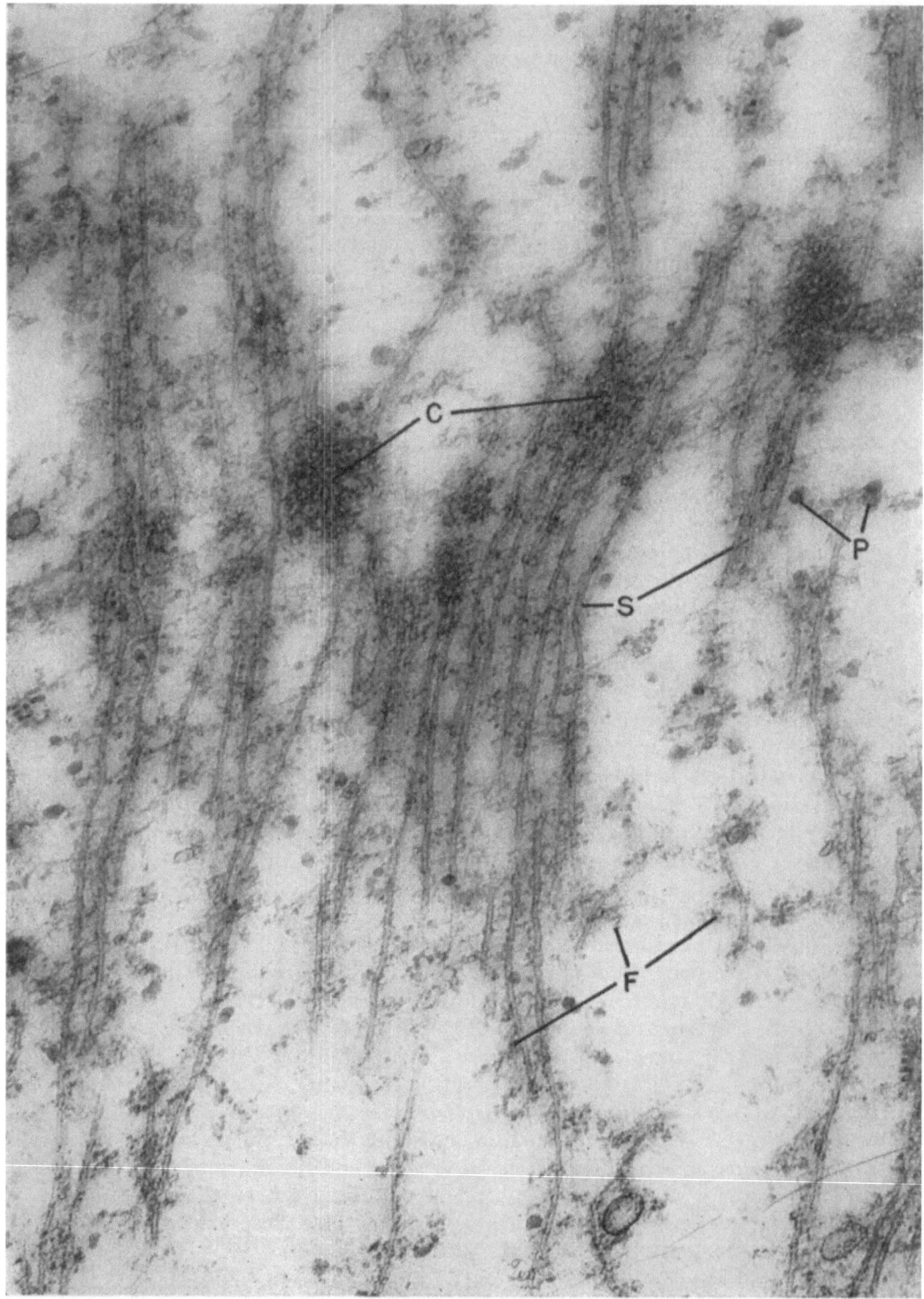

Fig. 21. Metaphase spindle of *Pelomyxa carolinensis* sectioned parallel to the spindle filaments (*S*) and perpendicular to the chromosome plate. Fine material (*F*) can be seen adhering to the spindle filaments and a few ribosome-like granules (*P*) are scattered about. These display no preferential association with the spindle filaments. Some spindle filaments appear to terminate on the chromosomes (*C*) while others clearly pass without interruption from one side of the metaphase plate to the other. All spindle filaments appear to be similar in structure. X 80,000. From ROTH and DANIELS, J. Cell Biol. **12**, 1962, 57.

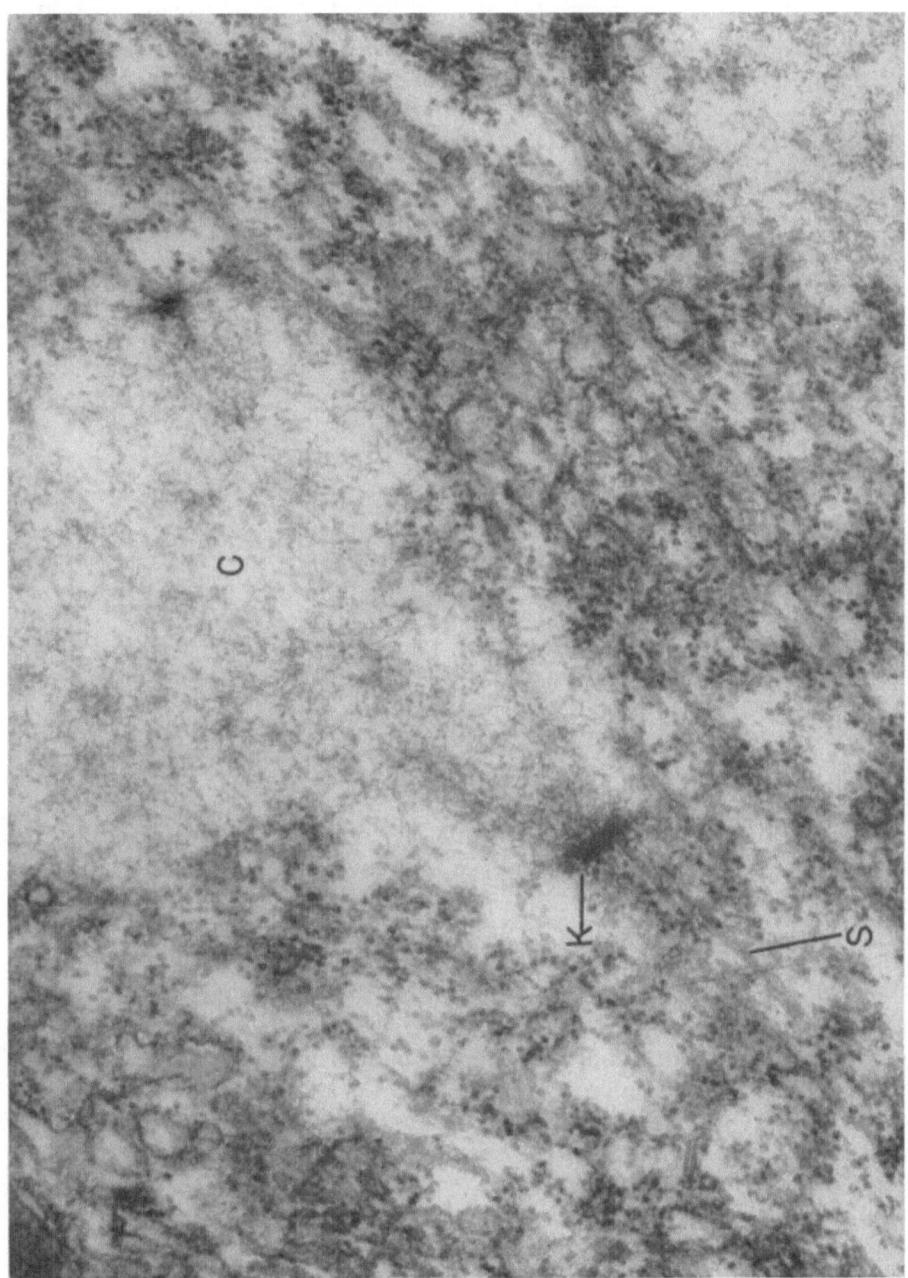

Fig. 22. This shows a chromosome at beginning anaphase of the first cleavage division in the sea urchin *Strongylocentrotus purpuratus*. The chromosome (C) is just beginning to split and its two kinetochores have started to move to opposite poles. The spindle filaments (S) have little or no fine material adhering to them. A number of spindle filaments can be seen to converge on the kinetochore. X 65,000. From HARRIS, J. Cell Biol. **14**, 1962, 475.

to find vesicles in both the spindle and astral regions which may in some cases be derived from the endoplasmic reticulum (Ruthmann, 1958; Porter and Machado, 1960; Harris, 1962 b; Buck, 1961). Electromicrographs of onion root tip cells in mitosis by Porter and Machado (1960) show the material of the spindle region arranged in whorl-like configurations around each chromosome. The endoplasmic reticulum did not intrude into this

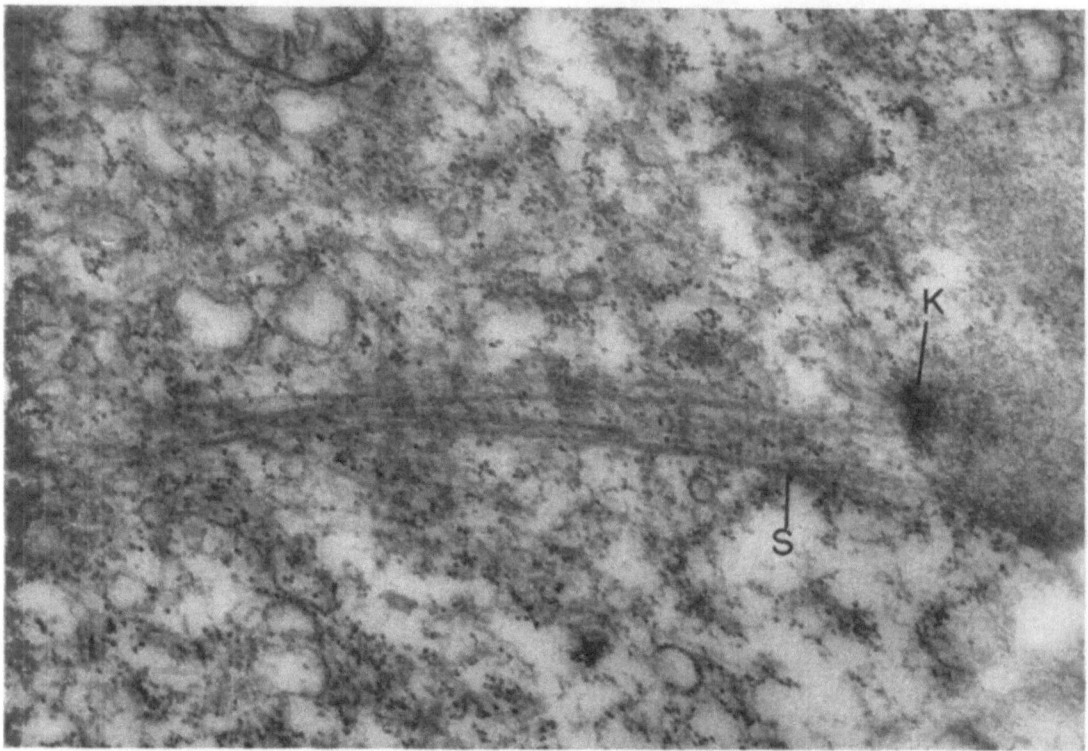

Fig. 23. This shows part of an anaphase figure in a 32-cell stage embryo of the sea urchin *Strongylocentrotus purpuratus*. Several spindle filaments emanating from the kinetochore (*K*) can be seen to merge with several continuous spindle filaments (*S*) to form a larger bundle. X 51,000. From Harris, J. Cell Biol. 14, 1962, 475.

region, but was seen to invade other regions of the spindle. No spindle fibers were evident in these photographs. In prometaphase and metaphase a concentration of endoplasmic reticulum appeared at the polar cap of the spindle. Not all electronmicrographs of spindle and asters reveal spindle filaments. Ito (1960) observed that the asters in the primary spermatocyte at metaphase of *Drosophila virilis* were made up of tubular or canalicular elements and that no real astral rays were visible; only structures described as astral lamallae. No spindle fibers were seen, although the spindle region was outlined by several layers of paired membranes. These membranes may correspond to the surface film described by Wada (1950) which separates the spindle body of *Tradescantia* from the remainder of the cytoplasm and which was postulated to be essential to the formation of the spindle. It has been postulated by Kurosumi (1957) and Kurosumi et al (1958) that the spindle and astral fibers may originate as linear chains of granules which fuse with each other to yield the tubular filaments.

Observations by HARRIS (1962 b) on the mitotic apparatus in sea urchin embryos substantiated earlier reports that the spindle fibers described by light microscopists were bundles of tubular elements (spindle filaments)

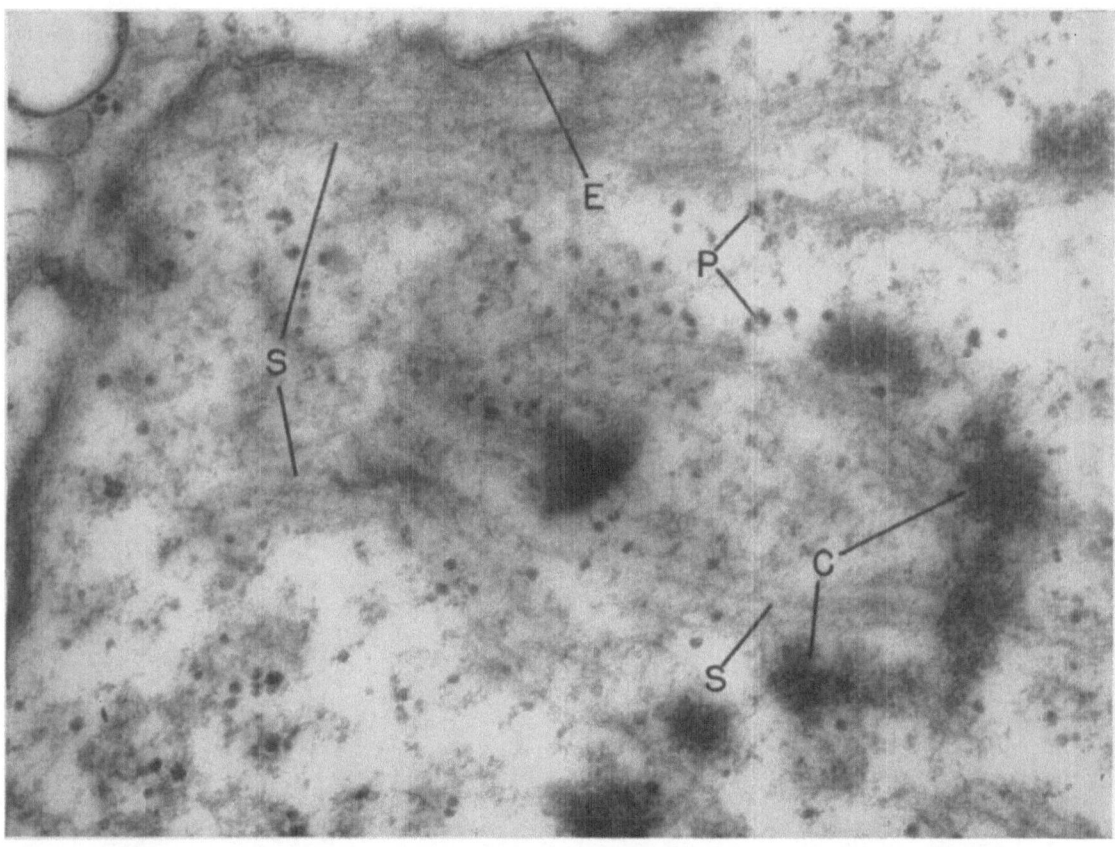

Fig. 24. Longitudinal section of early spindle filaments (S) and some chromosomes (C) in *Pelomyxa carolinensis*. Remnants of the nuclear envelope (E) are clearly visible. Numerous ribosome-like particles (P) are visible. Although these show no specific association with the spindle filaments, most of them seem to be surrounded by fine material. X 24,000. From ROTH and DANIELS, J. Cell Biol. **12**, 1962, 57.

about 15 mμ in diameter (Fig. 22 and 23). Of special interest was the observation that the spindle filaments emanating from the kinetochore were often seen to join a similar bundle of continuous fibers. The resultant configuration gave the impression that the chromosome was actually attached to the continuous fiber at a particular point and lends support to descriptions of similar configurations seen by light microscopists (see SCHRADER 1953). The asters appear to consist of a "matrix" of densely packed, irregularly oriented membrane structures near the center and elongated vesicles, predominantly radially oriented, near the periphery. Within the "matrix", astral filaments, chromosome-to-pole, and pole-to-pole fibers could be found, but mitochondria and yolk particles were excluded from the central regions of the aster.

A very consistent pattern in regard to the ends of the chromosomal fibers appears to prevail. One can consistently see these to be attached to or to terminate directly on the chromosomes at one end, while the poleward end only rarely appears attached to the centriole. The usual condition at the latter end is for the fiber to stop some distance from the centriole. So

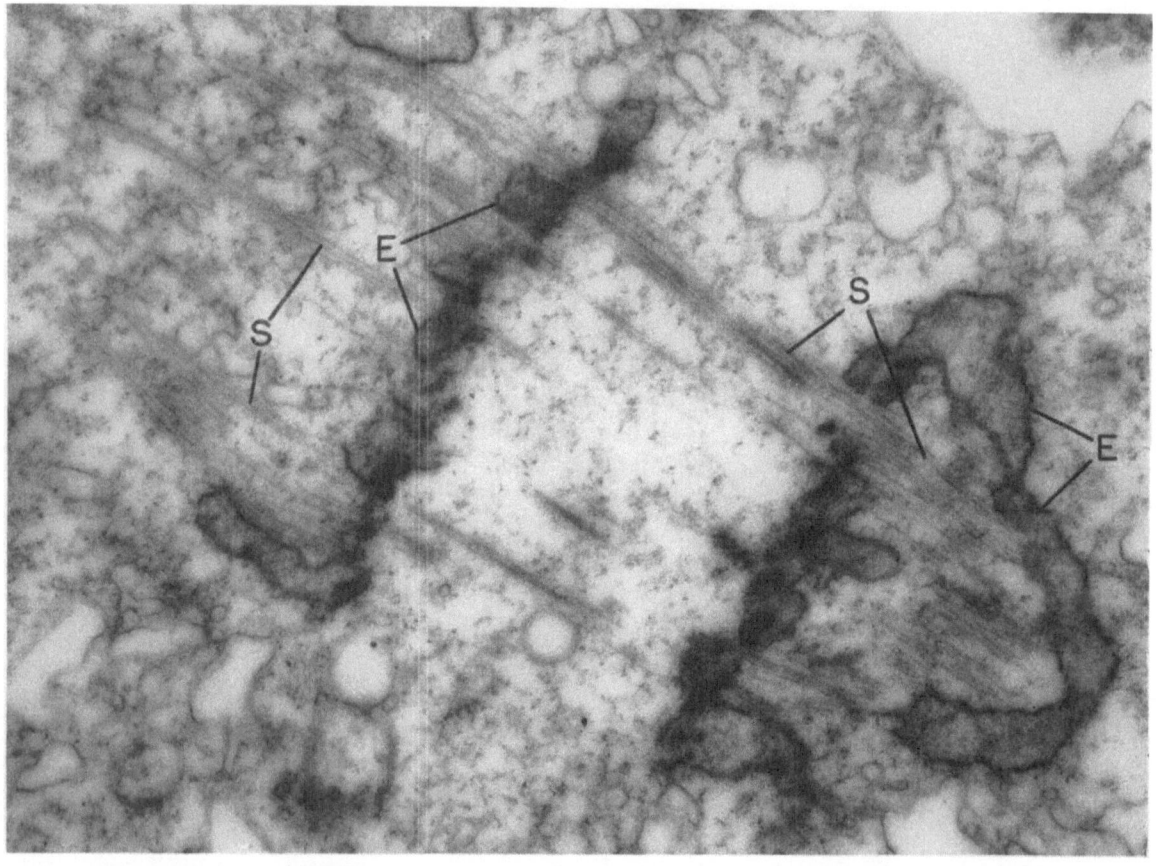

Fig. 25. An anaphase stage in *Pelomyxa carolinensis*. The material was fixed about 5 to 6 minutes after anaphase began and the chromosome plates are now separated by about 4 to 5 microns. Large remnants of the nuclear envelope (*E*) preceding the poleward movement of the chromosomes can be seen at the right. To the left other remnants of the envelope (*E*) are closely applied to the chromosomes. The spindle filaments (*S*) are surrounded by a layer of fine material. The continuous spindle filaments and the chromosomal spindle filaments appear structurally alike. X 11,000. From Roth and Daniels, J. Cell Biol. **12**, 1962, 57.

one cannot consider the spindle filaments to represent extensions of the tubes of the triplet fibers from which the centrioles are assembled; nor does any direct correlation exist between the number of spindle filaments and the tubes of the triplet fibers, for this would restrict the number of spindle filaments to a maximum of 27 per half spindle. Yet the centriole is clearly implicated in many cases as the entity guiding and organizing the assembly of the mitotic apparatus subunits into the completed structure. How the centrioles perform this function is not understood.

Turning our attention to the other end of the spindle filaments, we find that some completely bypass the chromosomes and that others terminate at localized, electron-dense loci on the chromosomes; i. e. kinetochores. The fact that we usually see a number of spindle filaments terminating on a single kinetochore can account for the spindle fibers which are visible with light microscopy. The bundle of spindle filaments, forming the spindle fiber and connected to a kinetochore, is sometimes a little more condensed immediately adjacent to the point of attachement than some distance away. In this connection it should be noted that in favorable electron micrographs there do not appear to be conspicuous differences along the length of the spindle fiber as one proceeds from the kinetochore toward the pole. Furthermore, the general impression gained from looking at a number of the published electronmicrographs of the component tubular filaments of the chromosomal, continuous and astral fibers is that they are very similar to each other and do not change in appearance through successive mitotic stages. (See figures 21, 24, and 25.) There is also little evidence for periodic cross-banding in the filaments (Roth and Daniels, 1962). The lack of any visible change in appearance of the chromosomal filaments in division stages is by no means fatal to the concept that these pull the chromosomes to the poles, through either contraction of the entire filament or a shortening achieved by removing material from either end. Harris (1962 b) suggests that perhaps the entire filament is in some manner drawn into the centrosphere, or that shortening of the filaments occurs at the kinetochore. She further states that the present evidence is not inconsistent with viewing the filaments as coil springs.

Some recent and excellent electronmicrographs by Szollosi (1964) may aid in improving our understanding of the structural relationship between the centriole and spindle fibers. These were of spermatid spindles and centrioles in the jellyfish *Phialidium gregarium*. The centrioles posses the structure typical for these bodies. Surrounding the distal centriole are satellites that are very similar to the pericentriolar bodies described by Bessis and Breton-Gorius (1958—see Figure 4). These satellites appear to be connected to the dense matrix between the tubular triplets of the distal centriole. Up to nine have been seen around one centriole. Satellites have not been demonstrated in association with the proximal centriole. Of greatest interest is the observation that tubular spermatid spindle fibers are seen to converge radially toward a satellite and fuse with the dense mass of this structure. At least one and possibly more spindle fibers may fuse with one satellite. It is perfectly clear in these electronmicrographs that the spindle fibers are not continuous with the tubular elements of the tubular triplets of the centrioles themselves. Since these observations are limited to a few meiotic and post-meiotic cells in a single species, caution must be exercised in extrapolating these observations to other species. Szollosi concludes that in young *Phialidium* spermatids, two distinct functions may be performed simultaneously by the centrioles: (1) formation of sperm tail or flagellum and (2) indirectly as attachment points for spindle fibers.

On the whole, electronmicroscopy has contributed little toward improving our understanding of h o w the spindle filaments are formed and the manner in which they effect chromosome movement.

The problem of whether the spindle fibers develop from the poles to the kinetochores or in the reverse direction has not been resolved and it is likely that either pattern may prevail, depending upon the material. In the flagellate protozoa so carefully studied by CLEVELAND (1957 b) astral rays grow out from each centriole. For a while they are free, but soon some from one centriole become long enough to meet others arising from the other centriole. As they meet, they join and elongate along one another to form the central spindle. The central spindle lengthens and begins to depress the ever intact nuclear membrane. At about the same time, some astral rays have become attached to the kinetochores embedded in the nuclear membrane and assume the role of chromosomal fibers. There is no apparent difference between the astral rays that form the central spindle and those which become the chromosomal fibers.

SATO (1960) has an interesting electronmicrograph which was interpreted as showing the very beginning of the filaments that are to form the chromosomal fiber. This shows the presence of a roughly linear array of granules originating from the kinetochore region. It is then imagined that these oriented granules fuse to form the filaments of the chromosomal fiber.

E. Evidence from other techniques

Irradiation of specific localized regions of living cells by a microbeam of ultraviolet light is another potentially very powerful technique for elucidating the function and time of greatest activity of specific subcellular structures. In this connection some recent studies by IZUTSU (1959, 1961 a, 1961 b) are of great interest. Selected parts of living grasshopper spermatocytes were irradiated with a U.V. microbeam two microns in diameter during the first meiotic division. When one centriole was irradiated at late diakinesis, the corresponding pole disappeared completely; the other pole behaved normally and a monocentric configuration of the division figure was observed. Following a prolonged metaphase, the chromosomes underwent apparently normal anaphase movement. Sometimes the irradiated pole nearly disappeared and then a new one appeared nearby. In this case most of the bivalents became oriented between the non-irradiated pole and the new pole. When one or both poles were irradiated at metaphase, the kinetochores did not modify their orientation with respect to the poles and these cells entered anaphase almost simultaneously with the non-exposed control cells. The kinetochores were also irradiated, and the subsequent behavior observed. When one kinetochore of a bivalent was irradiated at the end of metaphase I it remained in s i t u on the equator throughout anaphase, while the non-irradiated homolog moved toward its pole. When the kinetochore of a chromosome was irradiated during anaphase, the chromosomes continued to move to the pole in the company of those whose kinetochores had not been irradiated. It would be of great value to observe the behavior of a centriole during the division following the one in which it had been irradiated during anaphase.

In some Diptera, the *Cecidomyiidae*, chromosome elimination from somatic nuclei occurs as a regular feature during early cleavage. The exact division when this occurs is species dependent. In *Wachtliella persicariae* L. GEYER-DUSZYNSKA (1961) mentions that the only factor that inhibits the elimination of the E chromosomes and is also responsible for modifying an elimination mitosis into a normal division (one in which no chromosome

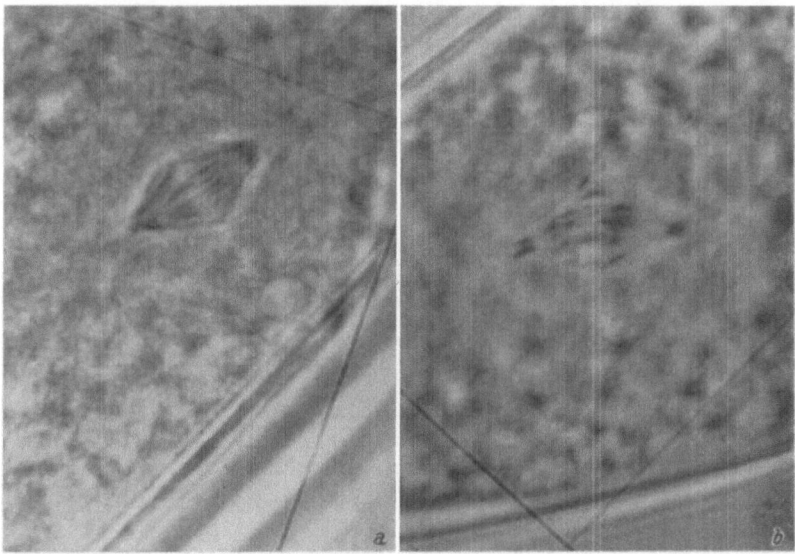

Fig. 26. Both (*a*) and (*b*) are photographs of normal mitotic divisions of somatic nuclei in untreated embryos. One can see chromosome elimination. From GEYER-DUSZYNSKA, Chromosoma (Berl.) **12**, 1961, 233—247.

elimination occurs) is a homogeneous substance that stains deeply with hematoxylin. In normal embryos this material is localized in the pole plasm which is the region from which the mother germ cell will arise. It is only in this cell and all of its descendants that no chromosome elimination occurs during cleavage. If this material is experimentally removed from the pole plasm, elimination of E chromosomes occurs in the nucleus of the mother germ cell. When this material is transplanted into the somatic part of the embryo, elimination of the E chromosomes does not occur in any nucleus present in the vicinity of this substance. It thus appears that this strongly hematoxylin positive material can transform an elimination division into a regular mitosis. GEYER-DUSZYNSKA (1961) has irradiated different regions of *Rhabdophaga batatas* Walsh embryos with ultraviolet microbeams and observed the subsequent effects upon appearance and shape of the cleavage spindles. The microbeam was focused on a part of the cytoplasm such that no nuclei were exposed directly to the ultraviolet microbeam, and the duration of exposure extended from interphase into prophase or metaphase of the division to be studied. The effects of such irradiation treatments were multiple: complete degeneration of nuclei in

the vicinity of irradiation; spindle disappearance and quasirosette formation; and multipolar mitosis of various configurations. Of interest here is
the observation that spindle disappearance follows the irradiation of regions
of the embryo some distance removed from the actual location of the
spindle. (See Figs. 26 and 27.) It is suggested by Geyer-Duszynska that
radiation induced changes in the cytoplasm adjacent to the spindle are
injurious to the centromeres of the chromosomes not eliminated (S), for it
is clear that the disappearance of the spindle fibers is not produced by
direct action of the ultraviolet upon the fibers themselves. From the experiments it is apparent that the behavior of the eliminated (E) chromosomes

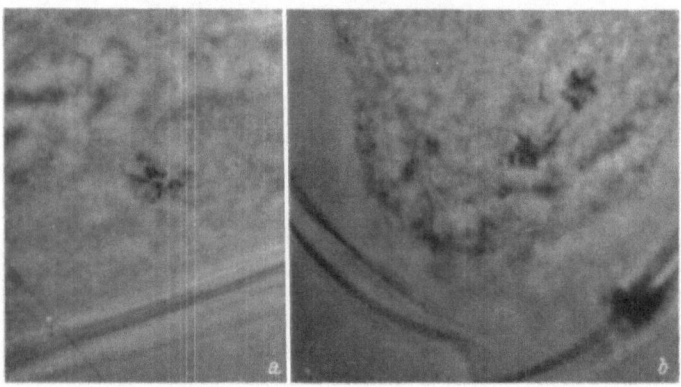

Fig. 27. Both (a) and (b) are from embryos irradiated in the fourth division. (a) shows a single quasi-rosette;
(b) shows daughter quasi-rosettes in the nucleus of a germ mother cell. There is no trace to be seen of the spindle
or asters. From Geyer-Duszynska, Chromosoma (Berl.) 12, 1961, 233—247.

and those (S) chromosomes (not normally eliminated) is the same in those
mitoses in which spindles are absent. The normal elimination of E chromosomes does not stem from the absence of chromosomal spindle fibers or
from intrinsically defective centromeres, but is hypothesized to be the result
of chemical changes in the adjacent cytoplasm that are injurious to the
centromeres of the E chromosomes.

Zirkle, Uretz and Haynes (1960) have also demonstrated spindle destruction in *Haemanthus* endosperm cells following ultraviolet microbeam
irradiation of a small portion of the cytoplasm of metaphase stages. They
used the disappearance of birefringence of the spindle fibers as the criterion
for the destruction of the spindle. When a small region of the cytoplasm
near one end of the cell was irradiated, by about 7 minutes after treatment
the birefringence of the entire spindle was nearly gone and by 14 minutes
spindle birefringence had entirely disappeared. Apparently spindle birefringence does not reappear, but the chromosomes in some cases undergo
a rather haphazard rearrangement into several distinct groups, each of
which becomes surrounded by a nuclear membrane. In similar studies on
newt (*Triturus*) cells in tissue culture spindle disappearance induced by
ultraviolet irradiation was followed by a very characteristic series of events
(Bloom, Zirkle and Uretz, 1955). First the metaphase chromosome configuration was completely lost and followed by the orderly rearrangement of the

chromosomes into a rosette with their kinetochores adjacent to a common attractive locus of unknown nature. Then the rosette separates into two daughter rosettes made up of undivided metaphase chromosomes. These two rosettes each form normal appearing nuclei which, however, must be highly aneuploid. The orderly progression of events following spindle disappearance in the newt might be attributable to the centrosomes which might not be seriously affected by cytoplasmic irradiation, although centrosomes have not been demonstrated at the centers of the rosettes. On this basis one would predict that, in view of the absence of centrosomes in the seed plants, there would be no orderly progression of events following spindle destruction on *Haemanthus* cells. This has been observed.

Ultraviolett microbeam irradiation of localized regions of the mitotic spindle itself, rather than a remote region of the cytoplasm, and the subsequent effects upon the birefringence of the spindle fibers have been performed by INOUÉ (1964) in collaboration with Dr. HIDEMI SATO. These experiments were done to test the concept that spindle fibers are formed in reference to orienting centers and that the spindle fibers themselves result from an orientation equilibrium with the unstructured region surrounding the individual fibers. If the action of the orienting center can be inhibited, the oriented material forming the spindle fiber should become disoriented and the fiber would loose its birefringence. On the other hand, direct experimental modification of a localized segment of the spindle fiber, without effecting the orienting center, should lead to the disorientation (and loss of birefringence) of the fiber distal to the "injury" while the material of the fiber proximal to the "injury" should retain its birefringence owing to the orienting influence of the unaffected center. These postulates have been experimentally verified. The experimental material used in this study (INOUÉ, 1964) was *Haemanthus* endosperm cells that had been somewhat flattened. When a cell in early anaphase was irradiated so that the microbeam extended across the basal regions of some chromosomal fibers and their kinetochores, the entire length of the chromosomal fibers including the distal unirradiated portions lost their birefringence. The birefringence did not reappear for a considerable period of time. When only some fibers and none of the corresponding kinetochores were irradiated, the birefrigence disappeared from the irradiated region and the portion distal to it, but did not disappear from the portion of the fibers between the kinetochores and the irradiated portion. Furthermore the birefringence of the irradiated and distal regions recovered in the course of a few minutes.

At this point an interesting comparison of the effect to the spindle can be made between microbeam irradiation of the cytoplasm and irradiation of a localized portion of the spindle itself (in *Haemanthus*). Irradiation of a portion of the cytoplasm remote to the spindle resulted in the apparently irreversible loss of birefringence by the entire spindle. Direct microbeam irradiation of the spindle fibers themselves resulted in the complete, but temporary, loss of birefringence of only the distal portions of the irradiated fibers. This suggests that two different mechanisms are involved in ultraviolet induced spindle disappearance depending upon the region irradiated.

Apparently the hypothetical precursor to the photochemically induced "spindle poison" (Zirkle, Uretz and Haynes, 1960) is restricted the cytoplasm.

Before discussing the very interesting experiments of Dietz (1959, 1962) on the role of kinetochores in spindle formation, it is pertinent to examine more closely other aspects of kinetochore-spindle pole associations and interactions. In view of what is known about chromosomes and kinetochores, combined with the preceding remarks on the behavior of centrioles, some discussion of an interesting and logically attractive hypothesis developed by Lettré and Lettré (1957, 1958, 1959) is relevant, although originally suggested by Resende (1947). A problem confronting any cell during division is that of insuring the movement of sister kinetochores, each with its attached chromatid, to opposite spindle poles. The Lettré and Lettré hypothesis states that the kinetochore is permanently attached by a fiber, which persists through interphase, to the center that will become a spindle pole. Thus the kinetochore-fiber-center complex would form a persistent unit and at the time of mitosis all parts divide simultaneously. (The term c e n t e r refers to a functional entity with no direct implications of its structural nature.) In this manner the potentially hazardous act of establishing connections between sister kinetochores and opposite poles each time the cell divides is obviated. Cytological evidence demonstrating the persistence of fibrous connections between mitotic centers and interphase nuclei is cited in support of their hypothesis, in addition to other cytological data.

The pleasing logic and simplicity of this hypothesis is very evident; however, there are some kinetochore-to-center relationships which are difficult to reconcile with it. It is clear that there must nevertheless exist a mechanism for establishing d e n o v o a connection between the kinetochore and the center prior to the first cleavage division in those organisms where the centers of the first cleavage spindle are derived entirely from o n e gamete (as is frequently observed among highly anisogamous organisms). The kinetochores must sever connections with the center whose continuity is interrupted upon termination of meiosis and establish new connections with the center contributed by the other gamete. So it is still necessary at least once in the lifetime (at the time of fertilization) of such organisms for the chromosomes contributed by one gamete to establish new kinetochore-to-center connections with the center contributed by the other gamete.

Those situations where each gamete appears to contribute one center of the first cleavage spindle (Conklin, 1902, 1904), pose yet another difficulty. Immediately following the second meiotic division the chromosomes of the ootid would still be attached to one center, and this relationship would be retained through fertilization. The chromosomes introduced by the sperm would remain attached to their corresponding center. In order to maintain a persistent center-to-kinetochore connection for each chromosome, and insure the movement of sister chromatids of both male and female chromosomes to opposite poles at the first cleavage division, each spindle center would have to contain contributions from both gametes. To achieve this one might expect to see four asters at some stage during the normal sequence of events between sperm penetration and the formation of the first cleavage spindle, since the center contributed by each gamete would have to divide, yielding a total of four centers (two of male origin and two of female origin) which would each form the focus of an aster. Subsequently the asters would pair, male with female, to form two spindle poles each containing center material derived from both gametes. This conflicts with CONKLIN's (1902, 1904) observations in *Crepidula*, although FOL (1891) has published a paper entitled the "Quadrille of the Centers" in which he describes the conjugation not only of the nuclei but the central bodies (centrioles?) as well. These results were later overthrown (WILSON, 1928: p. 439). Alternatively, one could envisage the formation of two amphiasters each containing only chromosomes and centers derived from one gamete, which would then fuse to form the first cleavage spindle. To the best of the Author's knowledge there is no evidence to support this alternative.

Another difficulty in applying the LETTRÉS' hypothesis arises when one compares the center-to-kinetochore relationship during the first meiotic division with that during mitosis. During mitosis the sister kinetochores of each chromosome migrate to o p p o s i t e (sister) spindle poles. This would require the attachment of sister kinetochores to sister centers. The duplication of the center-fiber-kinetochore reproductive unit prior to each mitotic division can nicely maintain this obligatory relationship indefinitely through successive mitoses. However, the center-to-kinetochore relationship during the first meiotic division is just the reverse of that found in mitosis, if one makes the following assumptions regarding the kinetochores and their alignment at metaphase: (1) the paired homologous kinetochores of each tetrad (or bivalent) lie on opposite sides of the equatorial plate and normally n e v e r go to the same spindle pole. (RHOADES, 1961); (2) each homologous kinetochores is functionally a single entity,[1] although structurally it may be double, i.e. composed of two sister kinetochores. (LIMA-DE-FARIA [1953, 1958] states that there is no fundamental difference in structure between the kinetochore at mitosis and first division meiosis on the basis of light microscopy). In order for homologous kinetochores to move to opposite poles, the sister kinetochores of each homolog must be connected to the same center (instead of to sister centers as is the case with mitosis). The situation is simplified somewhat if, on the other hand, one assumes that each homologous kinetochore is structurally a single entity (that is, it has not duplicated into two sister kinetochores) with but a single fiber connection to the center. However, this implies that the center must then be able to duplicate itself without the concomitant duplication of the remainder of the kinetochore-fiber-center unit.

There still remains at metaphase I the problem of forbidding homologous kinetochores from remaining attached to the same center, for in the mitotic divisions preceding the onset of meiosis a l l of the chromosomes possess fiber (or functional) attachments to b o t h centers at metaphase, but in the first meiotic metaphase each chromosome (dyad) of a tetrad is connected

---

[1] In the case of amphitelic orientation in Tipulids the kinetochores are functionally double (BAUER, DIETZ and RÖBBELEN 1961).

only to o n e center. This obligate change in the kinetochore-to-center relationship is clearly recognized by Lettré and Lettré (1959), but they present no discussion of the mechanism which can achieve this change. One can readily visualize a model system, entirely consistent with the Lettré hypothesis, that can achieve this. The model fulfills two important prerequisites: (1) it insures proper chromosome orientation in relation to the poles for either mitosis or meiosis and maintains persistent kinetochore-to-center attachements, (2) it can carry through both meiotic divisions with no synthesis or duplication of any of its components. Thus, prior to either a mitotic division or the first meiotic division, each "duplicated" center must be potentially able to determine 4 poles; that is, its multiplicity must correspond to the number of chromatids in a metaphase I bivalent. In this model the "duplicated" center can separate along either one of two planes that are perpendicular to each other. The plane along which separation will occur, which would coincide temporally with the migration of centrioles to poles, is determined by whether a mitotic or the first meiotic division is to follow. If a mitotic division follows, the duplicated center separates along the mitotic plane which means that sister kinetochores are attached to sister (opposite) poles. But if the first meiotic division is to follow, the duplicated center separates along the other plane (the meiotic plane) with the consequence that sister kinetochores of non-homologous chromosomes all remain attached to the same center and those of the homologous chromosome set are attached to the other center. This insures the movement of homologous chromosomes to opposite centers. The separation of the center at the second meiotic division is along the mitotic plane. The model is easily duplicated between successive mitotic divisions and the same configuration serves also as the starting point for meiosis. What the model cannot easily do (and this is not easily accounted for by the Lettré hypothesis) is to insure that the attachment of homologous chromosomes occurs randomly to opposite centers at meiosis as is required to account for the independent assortment of chromosomes.

The problem of re-establishing proper kinetochore-to-center attachments for the mitotic divisions following fertilization has been discussed. In those organisms (certain protists, mosses, algae, ferns), which are haploid for a considerable fraction of their life cycle, the transition from meiotic to mitotic division entails no changes in the kinetochore-to-center relationship extant upon conclusion of meiosis and calls for no change in the model discussed.

Dicentric chromosomes have been observed in some organisms, especially following X-irradiation. Usually such chromosomes are eliminated within a few generations; however, Sears and Câmara (1952) describe the behavior of a transmissible dicentric chromosome in wheat. This chromosome has a "primary" kinetochore which is apparently normal and a "secondary" kinetochore which is weaker than the primary but is nevertheless active in both mitosis and meiosis. Usually the dicentric chromosome is unpaired, so that at meiosis the two kinetochores oppose each other and both are active. That is, the kinetochores behave as though they were homologous—each being associated with a different pole. The secondary kinetochore is less strongly attracted to its pole than the primary one is, which can account for the dicentric univalent frequently going entirely to one pole at the first meiotic division. If the dicentric fails to divide during the first division and goes to one or the other pole, it usually divides normally in the second meiotic division. In some plants investigated, the cells were seen to contain a pair of dicentric chromosomes which at meiosis formed a bivalent. When the chromosomes paired, as a rule only one kinetochore of each was active and the behavior of the bivalent was normal. A very closely related problem was investiaged by DOYLE (1956), who studied the meiotic behavior of chromosomes during microsporogenesis in monoploid *Oenothera hookeri* ($n = 7$). A monoploid has only one genome and theoretically should show no pairing. However, pairing between what were considered non-homologous chromosomes was observed, but was thought to be the result of spiralizationpairing of two chromosomes which got too close to each other during pachytene. Observations on chromosome configuration in pollen mother cells at diakinesis revealed the presence of seven univalents in 59% of the cells, five univalents and one bivalent in 31% of the cells and other configurations in the remaining 10% of the cells. The bivalents behaved normally at the first meiotic division, while the univalents went to either one or the other of the two poles. Apparently DOYLE never observed the two chromatids of a univalent going to opposite poles during this division. He classified cells at anaphase I on the basis of the chromosome distribution between the two poles. This was seen to vary from 7–0 to 4–3. It was determined statistically that the chromosomes were not distributed randomly (on the basis of a random binomial distribution), and that the bias favored the 7–0 and 6–1 distributions. The deviation from a random binomial distribution was ascribed to: (1) disjunction of a bivalent (which was seen in 31% of the cells) and would give a bias toward the 6–1 distribution; (2) the failure of the spindle to separate the chromosomes into two groups, which would give a bias toward the 7–0 distribution. The magnitude of the bias introduced by these irregularities is not known. Therefore, it would be difficult to state with confidence that the affinity of any univalent for either pole was random (this, of course, is the case in meiosis in diploid cells). The second meiotic division was normal in that the chromatids of each univalent went to opposite poles, but there were anywhere from one to seven chromosomes in the spindle. DARLINGTON (1932) indicates that the behavior of univalents at the first meiotic division may follow one of the two courses: (1) if a univalent lies some distance from the equator, it will go entirely to the nearest pole; (2) if the univalent lies on or near the equator, it orients itself axially and separates into two chromatids which move to opposite poles. The second alternative is not easily reconciled with the LETTRÉ concept.

BAUER, DIETZ and RÖBBELEN (1961) present and discuss the results of an extensive investigation of kinetochore behavior inferred from observations on chromosome movement in translocation heterozygous *Tipula oleracea* spermatocytes. They used F_2 males obtained by mating F_1 sons of irradiated

P-males with normal females. Specifically, they selected for study those individuals bearing reciprocal translocations between an autosome and the small Y-chromosome. This material allowed them to observe the behavior of univalents, bivalents and trivalents within the same cell. Their results show that during the first maturation division the kinetochores of the two chromatids of each chromosome can be oriented in either of two ways. They may be oriented to the same pole (i.e. "syntelic") or they may be oriented to different poles (i.e. "amphitelic"). All chromosomes, whether univalent or part of a bivalent or trivalent, can behave in this manner. For example, in a bivalent: (1) the kinetochores of each dyad may be syntelic and both dyads directed toward the same pole; (2) the kinetochores of each dyad may be syntelic and each dyad directed toward different poles; (3) the kinetochores of one dyad may be syntelic and the other dyad may be amphitelic; and (4) the kinetochores of both dyads may be amphitelic. The frequencies of syntelic and amphitelic orientation of the kinetochores vary from 0–100% depending upon the position of the kinetochores in different meiotic pairing configurations (univalents, bi—and trivalents). Another interesting feature of kinetochore behavior is that syntelic kinetochores have been observed to change their orientation from one pole to another (see also Dietz, 1959). Thus the syntelic association with one pole is abolished and syntelic interaction with a new pole is established within one to three minutes. Such kinetochore reorientation may occur up to seven times before anaphase, but none occurs subsequent to its onset.

Observations show that (1) the sister kinetochores of a dyad are not invariably oriented toward the s a m e pole and (2) the kinetochores can switch allegiance from one pole to another. When this information is considered in context with what has already been said about kinetochores, it is clear that apparently no currently extant hypothesis can account for kinetochore-to-pole relationships.

The interesting observations of Dietz (1959) are also not readily reconciled with persistent kinetochore-to-center attachments. He can induce the formation of centrosomeless and multipolar spindles and has observed that some chromosomes can apparently change their allegiance from one pole to another.

Artificially activated sea urchin eggs are capable of producing structures similar to typical amphiasters of sperm-fertilized eggs (Dirksen, 1961 a). In these cells there did not seem to be much correlation between the centers and chromosomes. If the Lettré hypothesis were applicable, one would expect to observe that there would always be an equal number of chromosomes attached to each pole in those cases where the chromosomes were associated with more than one center. This clearly was not the case in all instances. One cannot include cytasters to which no chromosomes appear to be attached, since the egg may have contained potential centers in addition to the one or two to which the chromosomes would have to be attached prior to activation.

Thus certain fundamental aspects of chromosome behavior relative to the centers are not readily accounted for by a literal interpretation of the

LETTRÉ and LETTRÉ hypothesis. But the other extreme of the d e n o v o formation of kinetochore-to-center connections (functional and/or structural) at the time of each division also presents serious difficulties.

LIMA-DE-FARIA (1958) and DARLINGTON (1932) discuss other concepts relating kinetochore organization to chromosome movement.

There are observations of a different nature which clearly show that kinetochores can play a dominant role in organizing the mitotic apparatus. Very interesting experiments by DIETZ (1959, 1962) on the primary

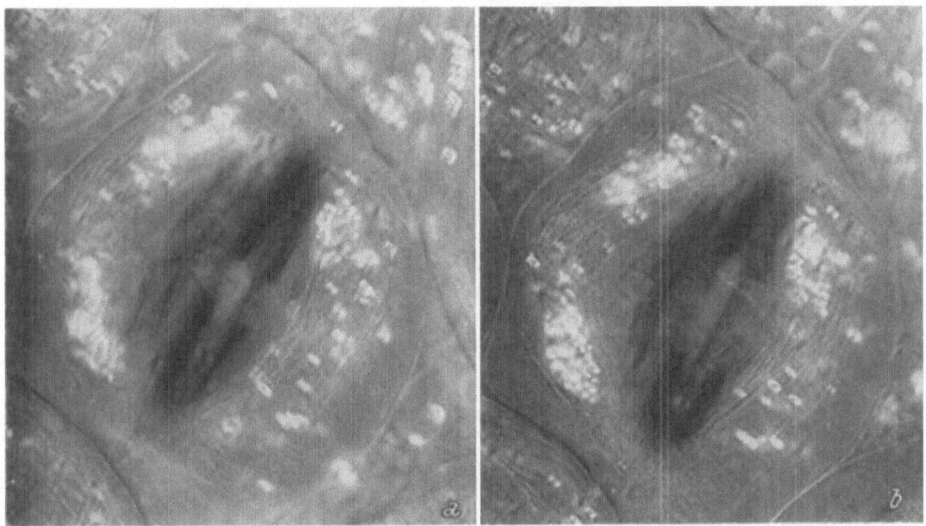

Fig. 28 *a* and *b*. Two stages in the normal first meiotic division in *Pales crocata* as seen with polarized light microscopy. This cell contains a normal bipolar, bicentric spindle. In Fig. 28 *a* is a metaphase stage in which the chromosomes can be seen as nearly clear vesicles at the equator. The chromosomes are joined to the poles by intensely birefrigent fibers which are uniformly birefringent along their length. In Fig. 28 *b* the cell has entered anaphase. The birefringence of the kinetochore-to-pole fibers is about the same seen in *a*. It is significant that the birefrigence of kinetochore-to-pole fibers terminates precisely at the chromosomes, while the zone between the separating chromosomes is free from birefringence. There are fibers of weaker birefringence running from pole-to-pole. Courtesy cf R. DIETZ.

spermatocytes of the crane-fly *Pales crocata* disclose that the normal relationship between spindle poles and the centrosome can be disturbed. In this connection, the principal observation is that centrosomes are not indispensible for the determination of a spindle pole. He has shown that bipolar spindles can be formed without the participation of one or both centrosomes. A chromosome-spindle normally appears in connection with each chromosome and completely independent of any influence that may be exerted by the centrosome. Initially the axes of the chromosome-spindles are randomly oriented, but later they become aligned parallel to each other without intervention from the centrosomes. This leads to the formation of a single bipolar spindle. (See Figure 28.) In Figure 29 one can see a bipolar spindle with one acentric pole. The other centrosome is not attached to the spindle and was not involved in its formation. Figure 30 shows a tripolar bicentric spindle. Flagella can be seen associated with the two centric poles. Monocentric bipolar spindles do not differ in size and shape from the normal bicentric spindles and chromosome movement in the two is the same. Dietz has also established that the acentric poles lack centrioles and that the centrosomes not connected to the spindle do have centrioles.

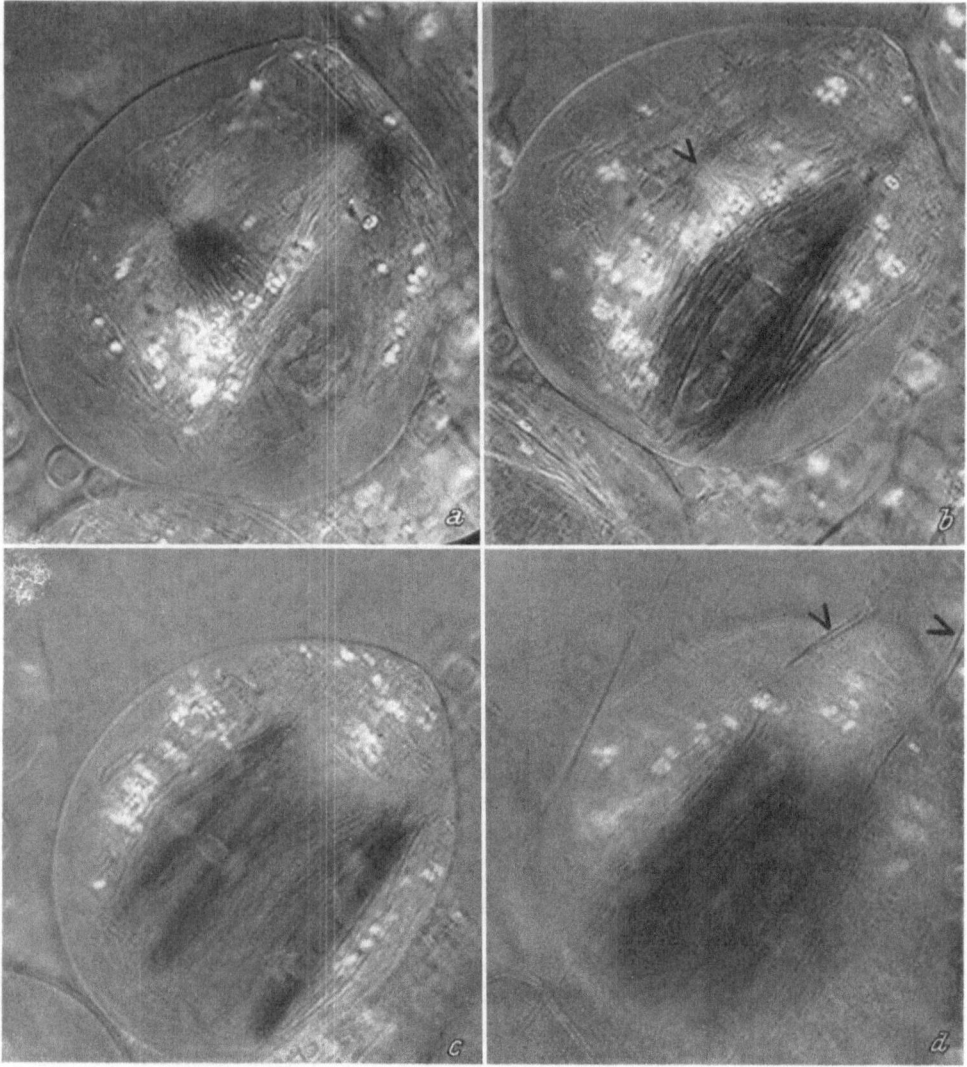

Fig. 29 a–d. Polarized light photomicrographs of a spermatocyte I of *Pales crocata* in which, by flattening, the asters were mechanically prevented from occupying opposite positions near the nuclear membrane in diakinesis. Fig. 29 a was taken 2 minutes after the disappearance of the nuclear membrane. The two birefringent asters can be seen. The chromosomes still lie in an isotropic zone. In Fig. 29 b, 23 minutes after breakdown of the nuclear membrane a positive birefringent bipolar spindle has formed around the chromosomes which have started their prometaphase movement with the first appearance of the spindle. One of the two asters is associated with the (upper) spindle pole, the other aster (arrow) lies apart from the spindle. Size and retardation of both asters have already decreased and there is no spindle between the two asters. The lower pole of the spindle is acentric. Fig. 29 c, 102 minutes after the disappearance of the nuclear membrane, the aster which was not attached to the spindle during a long period in early and mid prometaphase has moved to the upper pole and has become associated with it. Fig. 29 d represents the situation in early anaphase, 209 minutes after the disappearance of the nuclear membrane. The asters occupy about the same positions as in Fig. 29 c. Arrows identify paired flagellae associated with each aster. In spite of the fact that a spindle is formed with a bicentric and an acentric pole it differs from a normal spindle only by the fact that the spindle fibers in the pole regions are not focused. Chromosomal fibers of equal length, equal diameter and equal retardation connect the kinetochores with opposite spindle poles and movement of chromosomes is unaffected. Courtesy of R. Dietz.

In view of experimental evidence discussed above one may cautiously think of the spindle and asters being formed in relation to some orienting centers. It appears that the orienting centers for entire spindles and individual spindle fibers may be either kinetochores or centrioles, but there are too few data to permit a meaningful generalization. However the following summary appears valid:

1. In *Haemanthus* endosperm cells, in which no centrioles have been detected, kinetochores clearly orient material into chromosomal fibers.

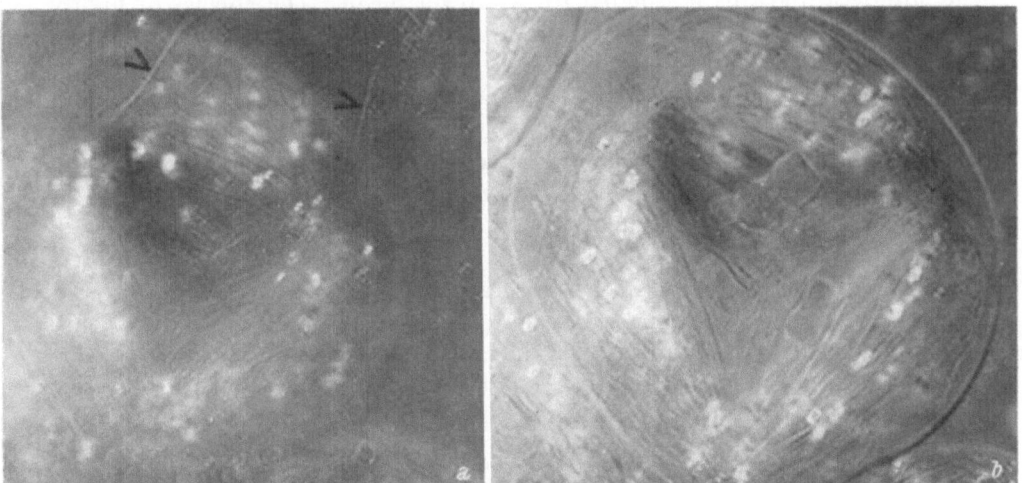

Fig. 30 *a* and *b*. Polarized light photomicrographs of an extremely flattened primary spermatocyte of *Pales crocata*. The same cell is seen in Figs. 30 *a* and *b*, and contains a tripolar bicentric spindle. The arrows in Fig. 30 *a* identify the flagellae. In Fig. 30 *b* the chromosomes can be seen to advantage. Courtesy of R. DIETZ.

2. In *Pales crocata* (DIETZ, 1959, 1962) spermatocytes, which do contain centrioles, spindle fibers and the polarity of the entire spindle can be established entirely independently of any centrioles that may be present, producing an acentric spindle pole. However, the asters present will always be formed in reference to a centriole, such that an acentric spindle pole will not have an aster.

3. In *Strongylocentrotus purpuratus* (MAZIA *et al*, 1960) it is possible to induce the formation of cells (in very early cleavage stages) that have half the normal number of mitotic centers (presumably centrioles). These cells try to produce a normal mitotic apparatus, but instead only succeed in forming a monopolar metaphase spindle that is lacking the spindle and aster on the side which is thought to have no centriole. It is not known whether or not in these cells the kinetochores have divided.

4. In *Barbulanympha* (CLEVELAND, 1957 b) astral rays grow out from each centriole. Soon some of those from opposite centrioles fuse to form the central spindle while others become chromosomal fibers by joining with the kinetochores embedded in the nuclear membrane. When only one pole was present (CLEVELAND, 1958), no chromosomes were ever connected to it, hence they did not move (compare with #3 above).

F. Summary

From the foregoing discussion, the following summarizing statements and conclusions about the structure and formation of the mitotic apparatus can be made:

1. The entire i n v i v o mitotic apparatus can be considered as a region made up of a visibly structured phase, which can be isolated as an intact body. It is bathed by an unstructured phase of randomly oriented molecules or micelles. Much of the unstructured phase can be expected to be lost when the mitotic apparatus is isolated by whatever means. The species of highly oriented subunits that impart the highly fibrous appearance to the mitotic apparatus region is also found randomly arrayed in the surrounding matrix. Presumably an equilibrium exists between the subunits in the oriented and non-oriented state.

2. The i n v i v o structural organization of the mitotic apparatus can be, and probably is, maintained by many weak bonds; but undoubtedly a very small number of intermolecular disulfide bonds (or other strong bonds) are also present.

3. The region of the mitotic apparatus contains many protein bound sulfhydryl groups which may alternately be reduced or oxidized during mitosis. The sulfhydryl groups of adjacent molecules are in close proximity to each other in the structured phase of the mitotic apparatus. The stability of the isolated mitotic apparatus can be ascribed to the oxidation of the protein-bound sulfhydryl groups into intermolecular disulfide bonds.

4. The normally fibrous organization of the mitotic apparatus can be disrupted i n v i v o (by colcemide and mercaptoethanol) and this region can then be isolated from the cell. Thus, the formation of the intermolecular bonds, which stabilize this region sufficiently to permit its successful isolation, can occur even when the molecules are randomly arrayed. This implies that the ability of the structural molecules of the mitotic apparatus to organize themselves into the normally fibrous configuration is not necessarily a property inherent in the molecules themselves (as in the case for collagen). Instead it may depend upon directions from specific entities (kinetochores and/or centrioles), although the structural molecules may possess the inherent property of mutual recognition (RNA may be involved in this property).

5. It is speculated that the functional activity of the mitotic apparatus reflects the making and breaking of bonds different from, and stronger than, those responsible for maintaining the structural integrity of the mitotic apparatus. The structured region of the mitotic apparatus is the mechanism by which it operates; the investing unstructured matrix provides the means for setting this mechanism into operation and maintaining it.

6. In addition to protein containing sulfhydryl groups, the mitotic apparatus region contains RNA, polysaccharides, lipoproteins and an ATPase. The contribution of these various other molecules to the structure and function of the mitotic apparatus is either not well understood or completely unknown.

7. The electron microscope has provided data indicating that the spindle fibers in different material may be constructed along the same or similar architectural patterns. A common feature seems to be, the tubular-appearing spindle filament which remains essentially unchanged during the course of mitosis. The continuous-fibers and chromosomal-fibers appear as bundles of spindle filaments. At the poleward end these are usually not seen to terminate directly on the centriole but some distance from it. The other end of the chromosomal-fiber, however, terminates directly on the kinetochore. Electron microscopy has improved our understanding of many of the structural aspects of mitotic apparatus, but has added little to our knowledge of how it functions.

8. In those cases in which the spindle fiber appears to develop from the kinetochore, the control mechanism which dictates when this event is to occur remains obscure, since it is reasonable to assume that the kinetochore retains its continuity from one cell generation to the next. However, whereas the centriole appears to remain structurally unchanged throughout the life cycle, we have no information about this aspect of the kinetochores. The control of centriole function, too, is not known.

Acknowledgement

The author takes this opportunity to acknowledge the generosity of Dr. BERNHARD and Dr. DIETZ in permitting me to use some of their previously unpublished photographs (Fig. 3–4, and Fig. 28–35 respectively). I also want to thank all the other investigators who gave me their permission to have some of their previously published figures reproduced in this article. These individuals and the specific references are acknowledged in the corresponding figure legends.

The author also thanks Academic Press, the publisher of Experimental Cell Research; the Rockefeller Institute, the publisher of the Journal of Cell Biology (formerly the Journal of Biophysical and Biochemical Cytology); La Semanie des Hopitaux, the publisher of Pathologie et Biologie; Springer-Verlag, the publisher of Chromosoma and Roux' Archiv für Entwicklungsmechanik der Organismen; and the American Association for the Advance of Science, the publisher of Science, for their permission to allow the reproduction in this article of figures that appeared im some of their publications.

The photographs in figures 21, 24 and 25 were taken at the Argonne National Laboratory.

I very gratefully acknowledge the suggestions and criticisms by Dr. DIETZ, Professor MAZIA and DR. P. HARRIS of various portions of the manuscript. These have contributed measureably to the final product.

The original work described in association with figure 20 was supported by grant No. GM 07286 awarded by the U.S. Department of Health, Education and Welfare.

Bibliography

AMANO, S., 1957: The structure of the centriole and spindle body as observed under the electron microscope and phase contrast microscopes. A new extension fiber theory concerning mitotic mechanisms in animal cells. Cytologia 22, 193—212.

AMBROSE, E. J., and A. BAJER, 1960: The analysis of mitoses in single living cells by interference microscopy. Proc. Roy. Soc. (Lond.) B. 153, 357—366.

AMOORE, J. E., 1963: Non identical mechanisms of mitotic arrest by respiratory inhibitors in pea root tips and sea urchin eggs. J. Cell Biol. 18, 555—567.

Bajer, A., 1958: Cine-micrographic studies on chromosome movements in beta-irradiated cells. Chromosoma (Berlin) 9, 319—331.
— 1961: A note on the behavior of spindle fibers at mitosis. Chromosoma (Berlin) 12, 64—71.
Bal, A. K., and P. R. Gross, 1964: Suppression of mitosis and macromolecular synthesis in onion roots by heavy water. J. Cell Biol. 23, 188—192.
Barthelmess, A., 1957: Chemisch induzierte multipolare Mitosen. Protoplasma 48, 546—561.
Bauer, H., R. Dietz, und C. Röbbelen, 1961: Die Spermatocytenteilungen der Tipuliden. III. Das Bewegungsverhalten der Chromosomen in Translokationsheterozygoten von Tipula oleracea. Chromosoma (Berlin) 12, 116—189.
Beams, H. W., T. C. Evans, V. van Breeman, and W. W. Baker, 1950: Electron microscopic studies on the structure of the mitotic figure. Proc. Soc. Exper. Biol. and Med. 74, 717—720.
Bělař, K., 1927: Beiträge zur Kenntnis des Mechanismus der indirekten Kernteilung. Naturwissenschaften 15, 725—734.
— 1929 a: Beiträge zur Kausalanalyse der Mitose. II. Untersuchungen an den Spermatocyten von Chorthippus (Stenobothrus) lineatus Panz. Arch. Entwicklungsmechanik d. Org. 118, 359—484.
— 1929 b: Beiträge zur Kausalanalyse der Mitose. III. Untersuchungen an den Staubfadenhaaren von Tradescantia virginica. Z. Zellforsch. 10, 73—134.
Bergen, P., 1960: On the blocking of mitosis by heat shock applied at different mitotic stages in the cleavage divisions of Trichogaster trichopterus var. sumatranus (Teleostei: Anabantidae). Nytt Mag. for Zool. 9, 37—121.
Bernhard, W., et E. de Harven, 1960: L'ultrastructure du centriole et d'autres éléments de l'appareil achromatique. Proc. 4th Intern. Conf. Electron Microscopy, 1958, (W. Bargmann, et al, eds.), Biologisch-Medizinischer Teil. Berlin: Springer-Verlag, vol. 2, 217—227.
Bessis, M., et J. Breton-Gorius, 1957 a: Étude au microscope électronique des granulations ferrugineuses des érythrocytes normaux et pathologiques. Rev. Hématol. 12, 43—63.
— — 1957 b: Le centriole des cellules du sang. Étude à l'état vivant et au microscope électronique. Bull. de Microsc. Appliquée 7, 54—56.
— — 1958: Sur une structure inframicroscopique péricentriolaire. Étude au microscope électronique sur les leucocytes de mammifères. C. R. Acad. Sci. (Paris) 246, 1289—1291.
— — et J. P. Thiery, 1958: Centriole, corps de Golgi et aster des leucocytes. Étude au microscope électronique. Rev. Hématol. 13, 363—386.
Bloom, W., R. E. Zirkle, and R. B. Uretz, 1955: Irradiation of parts of individual cells. III. Effects of chromosomal and extra-chromosomal irradiation on chromosome movements. Ann. N. Y. Acad. Sci. 59, 503—513.
Blum, H. F., and J. P. Price, 1950: Delay of cleavage in the Arbacia egg by U. V. radiation. J. Gen. Physiol. 33, 285—304.
Bonnevie, K., 1947: On the mechanics of mitosis. J. Morph. 81, 399—423.
Boss, J., 1955: Mitosis in cultures of newt tissues. IV. The cell surface in late anaphase and the movements of ribonucleoprotein. Exper. Cell Res. 8, 181—187.
Boveri, Th., 1888: Die Befruchtung und Teilung des Eies von Ascaris megalocephala. Zellen-Studien 2, Jena: Gustav Fischer.
— 1890: Über das Verhalten der chromatischen Kernsubstanz bei der Bildung der Richtungskörper und bei der Befruchtung. Zellen-Studien 3. Jena: Gustav Fischer.
— 1895: Über die Befruchtungs- und Entwicklungsfähigkeit kernloser Seeigel-Eier. Roux' Arch. Entwicklungsmech. (Berlin) 2. (Cited in Wilson 1928.)
— 1896: S. B. phys. Med. Ges. Würzburg 9, 133. (Cited in Dirksen 1961 a.)
— 1900: Über die Natur der Centrosomen. Zellen-Studien 4. Jena: Gustav Fischer.
Brachet, J., 1957: Biochemical Cytology. New York: Academic Press. xi/516.
Brauer, A., 1893: Zur Kenntnis des Spermatogenese von Ascaris megalocephala. Arch. mikr. Anat. 42. (Cited in Hertwig 1906, p. 196; Wilson 1928, p. 675.)
Briggs, R., and T. J. King, 1959: Nucleocytoplasmic interactions in eggs and embryos. Chapter 13 in: The Cell, vol. 1 (Brachet and Mirsky, eds.). New York: Academic Press. 537—617.
Bucher, N. L. R., and D. Mazia, 1960: DNA synthesis in relation to duplication of centers in dividing eggs of the sea urchin, Strongylocentrotus purpuratus. J. Biophys. Biochem. Cytol. 7, 651—655.

Bucher, O., 1959: Die Amitose der tierischen und menschlichen Zelle. Protoplasmatologia 6, 1—159.

Buck, R. C., 1961: Lamellae in the spindle of mitotic cells of Walker 256 Carcinoma. J. Biophys. Biochem. Cytol. 11, 227—236.

Burgos, M. H., and D. W. Fawcett, 1955: Studies on the fine structure of the mammalian testis. I. Differentiation of the spermatids in the cat (*Felix domestica*). J. Biophys. Biochem. Cytol. 1, 287—300.

— — 1956: An electron microscope study of spermatid differentiation in the toad, *Bufo arenarum hensel.* J. Biophys. Biochem. Cytol. 2, 223—240.

Carasso, N., 1958: Ultra-structure des cellules visuelles de larves d'amphibiens. Compt. rend. Sci. (Paris) 247, 527—531.

— and P. Favard, 1961: Les Ultrastructure Cytoplasmiques. 910—1117 in: Traité de microscopie électronique, vol. II, (C. Magnan, ed.) Paris: Hermann.

Carlson, J. G., 1940: Immediate effects of 230 R of X-rays on the different stages of mitosis in neuroblasts of the grasshopper, *Chortophaga viridifasciata.* J. Morphol. 66, 11—23.

— 1952: Microdissection studies of the dividing neuroblast of the grasshopper *Chortophaga viridifasciata* (De Geer). Chromosoma (Berlin) 5, 199—220.

— 1954: Immediate effects on division, morphology and viability of the cell. In: Radiation Biology, vol. 1, part 2 (A. Hollaender, ed.). New York: McGraw-Hill, 763—824.

Child, F. M., and D. Mazia, 1956: A method for the isolation of the parts of ciliates. Exper. 12, 161—162.

Clark, T. B., and F. G. Wallace, 1960: A comparative study of kinetoplast ultrastructure in the *Trypanosomatidae.* J. Protozool. 7, 115—124.

Cleveland, L. R., 1956: Brief accounts of the sexual cycles of the flagellates of *Cryptocercus.* J. Protozool. 3, 161—180.

— 1957 a: Types and life cycles of centrioles of flagellates. J. Protozool. 4, 230—241.

— 1957 b: Achromatic figure formation by multiple centrioles of *Barbulanympha.* J. Protozool. 4, 241—248.

— 1958: A factual analysis of chromosomal movement in *Barbulanympha.* J. Protozool. 5, 47—62.

— 1960 a: The centrioles of *Trichonympha* from termites and their functions in reproduction. J. Protozool. 7, 326—341.

— 1960 b: Photographs of living centrioles in resting cells of *Trichonympha collaris.* Arch. Protistenk. 105, 110—112.

— 1961: The centrioles of *Trichomonas* and their functions in cell reproduction. Arch. Protistenk. 105, 149—162.

— 1962: Photographs of gametogenesis in living cells of *Trichonympha.* Arch. Protistenk. 105, 497—508.

Coe, W. R., 1899: The maturation and fertilization of the egg of *Cerebratulus.* Zool. Jahrb. 12, 425—470.

Conklin, E. G., 1902: Karyokinesis and cytokinesis in the maturation, fertilization and cleavage of *Crepidula* and other Gasteropoda. J. Acad. Nat. Sci., Philadelphia, 2nd Ser. 12, 1—121.

— 1904: Experiments on the origin of the cleavage centrosomes. Biol. Bull. 7, 221—226.

— 1905: Organization and cell lineage of the ascidian egg. J. Acad. Nat. Sci., Philadelphia 13, 1—119.

Cornman, I., 1944: A summary of evidence in favor of the traction fiber in mitosis. Amer. Nat. 78, 410—422.

Costello, D. P., 1960: 183. The "polar suns" (centrospheres) of the egg of *Polychoerus* (*Turbellaria, Acoela*). Anat. Rec. 137, 346.

— 1961 a: On the orientation of centrioles in dividing cells and its significance: a new contribution to spindle mechanics. Biol. Bull. 120, 285—312.

— 1961 b: The orientation of centrioles in dividing cells and its significance. Biol. Bull. 121, 368.

Darlington, C. D., 1932: Recent Advances in Cytology. Philadelphia: P. Blakiston's Son and Co., Inc. xviii/559.

de Harven, E., and P. Dustin, Jr., 1960: Étude au microscope électronique de la stathmocinèse chez le rat. In: Les Actions antimitotiques et caryoclastiques des substances chimiques (J. Turchini, and P. Sentein, eds.) 189—198. Colloque No. 88, C. N. R. S., Paris.

Delamater, E. D., 1951: A new cytological basis for bacterial genetics. Cold Spr. Harb. Symp. Quant. Biol. 16, 381—409.

de Robertis, E., 1960: Some observations on the ultrastructure and morphogenesis of photoreceptors. J. Gen. Physiol. 43, (Suppl. 2), 1—14.

Dietz, R., 1958: Multiple Geschlechtschromosomen bei den cypriden Ostracoden, ihre Evolution und ihr Teilungsverhalten, Chromosoma (Berlin) 9, 359—440.
— 1959: Centrosomenfreie Spindelpole in Tipuliden-Spermatocyten, Z. Naturforsch. 14 b, 749—752.
— 1962: Polarisationsmikroskopische Befunde zur chromosomeninduzierten Spindel- bildung bei der Tipulide Pales crocata (Nematocera). Verh. Dtsch. Zool. Ges. Wien 1962. 131—138.

Dirksen, E. R., 1961 a: Studies on cytaster formation in artificially activated sea urchin eggs. Thesis (Ph. D. in Zoology), Univ. of Calif. ii, 152.
— 1961 b: The presence of centrioles in artificially activated sea urchin eggs. J. Biophys. Biochem. Cytol. 11, 244—247.
— 1964: The isolation and characterization of asters from artificially activated sea urchin eggs. Exper. Cell Res., 36, 256—269.

Doflein, F., and E. Reichenow, 1927—1929: Lehrbuch der Protozoenkunde. 5. Aufl. Jena: Gustav Fischer. viii, 1—1262.

Doyle, G. G., 1956: A comparative morphological and cytogenetic study of mono- ploid and autodiploid Oenothera hookeri, Thesis (Master of Science in Botany). Washington State University. vi, 29.

Eigsti, O. J., and P. Dustin, Jr., 1955: Colchicine — in Agriculture, Medicine, Biology and Chemistry. Ames: Iowa State College Press. xiii, 470.

Epel, D., 1963: The effects of carbon monoxide inhibition on ATP level and rates of mitosis in the sea urchin egg. J. Cell Biol. 17, 315—319.

Errera, M., and M. Brunfant, 1964: Observations of mitotic figures in pulse labelled HeLa cells. Exper. Cell Res. 33, 105—111.

Fankhauser, G., 1932: Cytological studies on egg fragments of the salamander Triton. II. The history of the supernumerary sperm nuclei in normal fertilization and cleavage of fragments containing the egg nucleus. J. Exper. Zool. 62, 185—235.
— 1948: The organization of the amphibian egg during fertilization and cleavage. Ann. N. Y. Acad. Sci. 49, 684—708.

Fawcett, D. W., 1958: The structure of the mammalian spermatozoon. Int. Rev. Cytol. 7, 195—234.
— 1961: Cilia and flagella, in: The Cell (J. Brachet, and A. E. Mirsky, eds.), vol. 2, 217—298.

Fetner, R. H., and E. D. Porter, 1965: Multipolar mitosis in the KB (Eagle) human cell line and its increased frequency as a function of 250 KV X-irradiation. Exper. Cell Res. 37, 429—439.

Fischer, A., 1899: Fixierung, Färbung und Bau des Protoplasmas. Jena: Gustav Fischer. x, 362.

Fol, H., 1891: Die „Centrenquadrille", eine neue Episode aus der Befruchtungs- geschichte. Arch. Sci. Phys. et Nat. 25, 393—420. (Cited in Dirksen 1961 a.)

Fry, H. J., 1929: The so-called central bodies in fertilized Echinarachnius eggs. I. The relationship between central bodies and astral structure as modified by various mitotic phases. Biol. Bull. 56, 101—128.
— 1932: Studies of the mitotic figure. I. Chaetopterus: Central body structure at metaphase, first cleavage, after picro-acetic fixation. Biol. Bull. 63, 149—186.

Gall, J. G., 1961: Centriole replication. A study of spermatogenesis in the snail Viviparus. J. Biophys. Biochem. Cytol. 10, 163—193.

Galtsoff, P. S., and D. E. Philpott, 1960: Ultrastructure of the spermatozoan of the oyster, Crassostrea virginica. J. Ultrastructure Res. 3, 241—253.

Gatenby, J. B., 1959: Electron microscopy of spermatogenesis of Lumbricus. Trans. Roy. Soc. N. Z. 1, 55.
— 1961: The electron microscopy of centriole flagellum and cilium. J. Roy. Micr. Soc. 79, 299—317.

Gaulden, M. E., and J. G. Carlson, 1951: Cytological effects of colchicine on the grasshopper neuroblast in vitro with special reference to the origin of the spindle. Exper. Cell Res. 2, 416—433.

Geitler, L., 1934: Grundriß der Cytologie. Berlin: Verlag von Gebrüder Borntraeger. viii, 1—295.

GEYER-DUSZYNSKA, I., 1961: Spindle disappearance and chromosome behavior after partial-embryo irradiation in *Cecidomyiida* (*Diptera*). Chromosoma (Berlin) 12. 233—247.

GIBBONS, I. R., 1961: Structural asymmetry in cilia and flagella. Nature (London) 190, 1128—1129.

— and A. V. GRIMSTONE, 1960: On flagellar structure in certain flagellates. J. Biophys. Biochem. Cytol. 7, 697—716.

GIESE, A., 1949: Activation of eggs, fertilization and early development as affected by ultraviolet rays. Amer. Nat. 83, 165—178.

GRASSO, J. A., H. SWIFT, and G. A. ACKERMAN, 1962: Obvervations on the development of erythrocytes in mammalian fetal liver. J. Cell Biol. 14, 235—254.

GRIFFIN, B. B., 1896: The history of the archoplasmic structures in the maturation and fertilization of *Thalassema*. Trans. N. Y. Acad. Sci. 2, 163—176.

GRIMSTONE, A. V., 1961: Fine structure and morphogenesis in Protozoa. Biol. Rev. 36. 97—150.

GROSS, P. R., and G. H. COUSINEAU, 1963: Synthesis of spindle-associated proteins in early cleavage. J. Cell Biol. 19, 260—265.

— L. I. MALKIN, and W. A. MOYER, 1964: Templates for the first proteins of embryonic development. Proc. Nat. Acad. Sci. (Wash.) 51, 407—414.

— D. E. PHILPOTT, and S. NASS. 1958: The fine structure of the mitotic spindle in sea urchin eggs. J. Ultrastructure Res. 2, 55—72.

— and W. SPINDEL, 1960 a: Heavy water inhibition of cell division: an approach to mechanism. Ann. N. Y. Acad. Sci. 90, 500—522.

— — 1960 b: Mitotic arrest by deuterium oxide. Science 131, 37—39.

HARRIS, P. J., 1962 a: The fine structure of the mitotic apparatus in dividing sea urchin blastomeres. Thesis (Ph. D. in Zoology), Univ. of California. iii, 95.

— 1962 b: Some structural and functional aspects of the mitotic apparatus in sea urchin embryos. J. Cell Biol. 14, 475—487.

HARVEY, E. B., 1936: Parthenogenetic merogony or cleavage without nuclei in *Arbacia punctulata*. Biol. Bull. 71, 101—121.

— 1940: A comparison of the development of nucleate and non-nucleate eggs of *Arbacia punctulata*. Biol. Bull. 79, 166—187.

HASHIMOTO, T., S. F. CONTI. and H. B. NAYLOR, 1959: Studies of the fine structure of microorganisms. IV. Observations on budding *Saccharomyces cerevisiae* by light and electron microscopy, J. Bacteriol. 77, 344—354.

HAY, E., 1962: Cytological studies of dedifferentiation and differentiation in regenerating amphibia limbs, in: Regeneration (D. RUDNICK, ed.), New York: Ronald Press Co. v, 1—272.

HEIDENHAIN, M., 1907: Plasma und Zelle, vol. I. Jena: GUSTAV FISCHER. viii, 1—506.

HENNEGUY, L. F., 1891: Nouvelles recherches sur la division cellulaire indirecte. J. de l'Anat. 27. (Cited in HERTWIG 1906, 219.)

— 1897: Sur les rapports des cils vibratiles avec les centrosomes. Arch. d'Anat. Microsc. 1, 481—496.

HENSHAW, P. S., 1940 a: Further studies on the action of roentgen rays on the gametes of *Arbacia punctulata*. I. Delay in cell division caused by exposure of sperm to roentgen rays, Am. J. Roentgenol. Radium Therapy 43, 899—906.

— 1940 b: Further studies on the action of roentgen rays on the gametes of *Arbacia punctulata*. II. Modification of the mitotic time schedule in the eggs by exposure of the gametes to roentgen rays, Amer. J. Roentgenol. Radium Therapy 43, 907—912.

— 1940 c: Further studies on the action of roentgen rays on the gametes of *Arbacia punctulata*. III. Fixation of irradiation effect by fertilization in eggs. Amer. J. Roentgenol. Radium Therapy 43, 913—916.

— 1940 d: Further studies on the action of roentgen rays on the gametes of *Arbacia punctulata*. V. The influence of low temperature on recovery from roentgen-ray effects in the eggs. Amer J. Roentgenol. Radium Therapy 43, 921—922.

— 1940 e: Further studies on the action of roentgen rays on the gametes of *Arbacia punctulata*. VI. Production of multipolar cleavage in the eggs by exposure of the gametes to roentgen rays. Amer. J. Roentgenol. Radium Therapy 43, 923—933.

— and I. COHEN, 1940: Further studies on the action of roentgen rays on the gametes of *Arbacia punctulata*. IV. Canges in radiosensitivity during the first cleavage cycle. Amer. J. Roentgenol. Radium Therapy 43, 917—920.

HERTWIG, O., 1893: Die Zelle und die Gewebe. Jena: GUSTAV FISCHER. xi, 1—296.
— 1906: Allgemeine Biologie. Jena: GUSTAV FISCHER. xvi, 1—649.
HOFFMANN-BERLING, H., 1954 a: Adenosintriphosphat als Beitreibsstoff von Zellbewe-
    gungen. Biochim. Biophys. Acta 14, 182—194.
— 1954 b: Die Bedeutung des Adenosintriphosphat für die Zell- und Kernteilungs-
    bewegungen in der Anaphase. Biochim. Biophys. Acta 15, 226—236.
HOLTZER, H., J. ABBOTT, and M. W. CAVANAUGH, 1959: Some properties of embryonic
    and cardiac myoblasts. Exper. Cell Res. 16, 595—601.
HUETTNER, A. F., 1933: Continuity of the centrioles in Drosophila melanogaster.
    Z. Zellforsch. 19, 119—134.
HUGHES, A., 1952: The mitotic cycle: The cytoplasm and nucleus during interphase
    and mitosis. New York: Academic Press, Inc. viii, 1—232.
HULTIN, T., 1961 a: The effect of puromycin on protein metabolism and cell division
    in fertilized sea urchin eggs. Exper. 17, 410—411.
— 1961 b: Activation of ribosomes in sea urchin eggs in response to fertilization.
    Exper. Cell Res. 25, 405—417.
— 1964: Factors influencing polyribosome formation i n v i v o. Exper. Cell Res. 34,
    608—611.
IMMERS, J., 1957: Cytochemical studies of fertilization and first mitosis of the sea
    urchin egg. Exper. Cell Res. 12, 145—153.
INOUÉ, S., 1959: Motility of cilia and the mechanism of mitosis. Rev. Mod. Phys. 31,
    402—408.
— 1960: On the physical properties of the mitotic spindle. Ann. N. Y. Acad. Sci. 90,
    529—530.
— 1964: Organization and function of the mitotic spindle, in: Primitive Motile
    Systems in Cell Biology, New York: Academic Press Inc., 549—598.
— and A. BAJER, 1961: Birefringence in endosperm mitosis. Chromosoma (Berlin) 12,
    48—63.
ITO, S., 1960: The lamellar systems of cytoplasmic membranes in dividing spermato-
    genic cells of Drosophila virilis. J. Biophys. Biochem. Cytol. 7, 433—442.
IZUTSU, K., 1959: Irradiation of parts of single mitotic apparatus in grasshopper
    spermatocytes with an ultraviolet-microbeam. Mie Medical Journ. 9, 15—29.
— 1961 a: Effects of U. V. microbeam irradiation upon division in grasshopper
    spermatocytes. I. Results of irradiation during prophase and prometaphase.
    Mie Medical Journ. 11, 199—212.
— 1961 b: Effects of U. V. microbeam irradiation upon division in grasshopper
    spermatocytes. II. Results of irradiation during metaphase and anaphase. Mie
    Medical Journ. 11, 213—232.
JOHNSON, H. H., 1931: Centrioles and other cytoplasmic components of the male
    sperm cells of the Gryllidae. Z. wiss. Zool. 140, 115—166.
KANE, R. E., 1962: The mitotic apparatus: isolation by controlled pH. J. Cell Biol. 12,
    47—55.
KAWAMURA, K., 1955: The course of spindle formation in the spermatocyte of the
    grasshopper, Acrydium japonicus, observed by phase microscopy. Cytologia
    (Tokyo) 20, 47—51.
— 1960: Studies on cytokinesis in neuroblasts of the grasshopper, Chortophaga
    viridifasciata (DE GEER). II. The role of the mitotic apparatus in cytokinesis.
    Exper. Cell Res. 21, 9—18.
KAWAMURA, N., 1960: Cytochemical and quantitative study of proteinbound sulf-
    hydryl and disulfide groups in eggs of Arbacia during the first cleavage. Exper.
    Cell Res. 20, 127—138.
— and K. DAN, 1958: A cytochemical study of the sulfhydryl groups of sea urchin
    eggs during the first cleavage. J. Biophys. Biochem. Cytol. 4, 615—620.
KELLY, D. E., 1962: Pineal organs: photoreception, secretion and development. Amer.
    Sci. 50, 597—625.
KERR, N. S., 1960: Flagella formation by myxamoebae of the true slime mold,
    Didymium nigripes. J. Protozool. 7, 103—108.
KING, H. D., 1901: The maturation and fertilization of the egg of Bufo lentiginosus.
    J. Morph. 17, 293—350.
KUPKA, E., und F. SEELICH, 1948: Die anaphasische Chromosomenbewegung. Ein
    Beitrag zur Theorie der Mitose. Chromosoma (Berlin) 3, 302—327.
KUROSUMI, K., 1957: Electron microscope studies on mitosis in sea-urchin blasto-
    meres. Protoplasma 49, 116—139.

KUROSUMI, K., M. YAMAGISHI, and T. NAGAKAWA, 1958: Electron microscopic and cyto-chemical studies on the cytoplasmic RNA in sea-urchin eggs. Okajimas Fol. Anat. Japon. **30**, 369—387.

LENHOSSEK, M., 1898: Über Flimmerzellen. Verh. dtsch. anat Ges., Jena: GUSTAV FISCHER, **12**, 106—128.

LEPPER, R., JR., 1956: The plant centrosome and the centrosome-blepharoplast homo-logy. Bot. Rev. **22**, 375—417.

LETTRÉ, H., und R. LETTRÉ, 1957: Persistenz der Chromosomenspindelfaser, eine Arbeitshypothese zur Deutung der karyokinetischen Vorgänge. Naturwiss. **44**, 406.

— — 1958: Un problème cytologique: La persistance des structures du fuseau dans l'intervalle des mitoses. Rev. Hémat. **13**, 337—362.

— — 1959: A cytological problem: permanence of the chromosomal spindle fiber during interphase. Nucleus (Calcutta) **2**, 23—44.

LEWIS, A. G., and G. MARIN, 1963: Induction of multipolar spindles by X-radiation in mammalian cells in vitro. Exper. Cell Res. **31**, 448—451.

LEWIS, M. R., 1933: Reversible changes in the nature of the mitotic spindle brought in living cells by means of heat. Arch. exper. Zellforsch. **14**, 464—470.

LILLIE, F. R., 1897: On the origin of the centres of the first cleavage-spindle in *Unio complanata*. Sci. N. S. **5**, 389—390.

LIMA-DE-FARIA, A., 1956: The role of the kinetochore in chromosome organization. Hereditas (Lund) **42**, 85—160.

— 1958: Recent advances in the study of the kinetochore. Int. Rev. Cytol. **7**, 123—158.

LINDEGREN, C. C., M. A. WILLIAMS, and D. O. McCLARY, 1956: The distribution of chromatin in budding yeast cells. Antonie van Leeuwenhoek **22**, 1—20.

LOEB, J., 1892: Investigations in physiological morphology. J. Morph. **7**, 253—262.

LONGWELL, A., and M. MOTA, 1960: The distribution of cellular matter during meiosis. Endeavour **19**, 100—107.

LORCH, I. J., 1952: Enucleation of sea-urchin blastomeres with or without removal of asters. Quart. J. Microsc. Sci. **93**, 475—486.

LWOFF, A., 1949: Kinetosomes and the development of ciliates. Growth. Symposium **9**, 61—91.

— — 1950: Problems in the morphogenesis of ciliates. New York: John Wiley and Sons, Inc. ix, 1—103.

McCLENDON, J. F., 1908: The segmentation of eggs of *Asterias farbesii* deprived of chromatin. Roux' Arch. Entwicklungsmechanik (Berlin) **26**, 662—668.

MAGGIO, R., and C. CATALANO, 1963: Activation of amino acids during sea urchin development. Arch. Biochem. Biophys. **103**, 164—168.

MANGAN, J., T. MIKI-NOUMURA, and P. J. GROSS, 1965: Protein synthesis and the mitotic apparatus. Science **147**, 1575—1578.

MARSLAND, D., and A. M. ZIMMERMAN, 1963: Cell division: differential effects of heavy water upon the mechanisms of cytokinesis and karyokinesis in the eggs of *Arbacia punctulata*. Exper. Cell Res. **30**, 23—35.

MAZIA, D., 1955: The organization of the mitotic apparatus. Symp. Soc. Exper. Biol. **9**, 335—357.

— 1957: Some problems in the chemistry of mitosis, in: Chemical Basis of Heredity (W. D. McELROY, and B. GLASS, eds.), Baltimore: The Johns Hopkins Press. 1—169.

— 1961: Mitosis and the physiology of cell division, in: The Cell, vol. III (J. BRACHET, and A. E. MIRSKY, eds), New York: Academic Press, 77—412.

— 1963: Synthetic activities leading to mitosis. J. Cell. Comp. Physiol. **62**, Suppl. 1, 123—140.

— R. CHAFFEE, and R. M. IVERSON, 1961: Adenosine triphosphate in the mitotic apparatus. Proc. Nat. Acad. Sci. (Wash.) **47**, 788—790.

— and K. DAN, 1952: The isolation and biochemical characterization of the mitotic apparatus of dividing cells. Proc. Nat. Acad. Sci. (Wash.) **28**, 826—838.

— P. J. HARRIS, and T. BIBRING, 1960: The multiplicity of the mitotic centers and the time-course of their duplication and separation. J. Biophys. Biochem. Cytol. **7**, 1—20.

— J. M. MITCHISON, H. MEDINA, and P. HARRIS, 1961: The direct isolation of the mitotic apparatus. J. Biophys. Biochem. Cytol. **10**, 467—474.

— and J. D. ROSLANSKY, 1956: The quantitative relations between total cell proteins and the proteins of the mitotic apparatus. Protoplasma **46**, 528—534.

Mazia, D., and A. M. Zimmerman, 1958: SH compounds in mitosis. II. The effect of mercaptoethanol on the structure of the mitotic apparatus in sea urchin eggs. Exper. Cell Res. 15, 138—153.
Mead, A. D., 1898: The origin and behavior of the centrosomes in the annelid egg. J. Morph. 14, 181—218.
Meves, F., 1903: Über Oligopyrene und Apyrene Spermien und über ihre Entstehung, nach Beobachtungen an *Paludina* und *Pygaera.* Arch. mikr. Anat. 61, 1—84.
Miki, T., 1962: The ATP-ase activity of the mitotic apparatus of the sea urchin egg. Exper. Cell Res. 29, 92—101.
— 1964: ATP-ase staining of sea urchin eggs during the first cleavage. Embryol. 8, 158—165.
Minouchi, O., 1936: Cytologische Studien über das Ei von *Polystomum integerrimum* von der Eiablage bis zu den frühen Furchungsstadien. Z. Zellforsch. 24, 85—127.
Mitchison, J. M., and M. M. Swann, 1953: Measurements on sea-urchin eggs with an interference microscope. Quart. J. microsc. Sci. 94, 381—389.
Monroy, A., and A. Tyler, 1963: Formation of active ribosomal aggregates (polysomes) upon fertilization and development of sea urchin eggs. Arch. Biochem. and Biophys. 103, 431—435.
Montgomery, P. O'B., and W. A. Bonner, 1959: U. V. time lapse motion picture observations of mitosis in newt cells. Exper. Cell Res. 17, 378—384.
Moore, A. R., 1938: Segregation of "Cleavage-substance" in the unfertilized egg of *Dendraster excentricus.* Proc. Soc. Exper. Biol. Med. 38, 162—163.
Morgan, T. H., 1896: The production of artificial astrophers. Arch. Entw. 3, 339—361.
Mundkur, B. D., 1954: The nucleus of *Saccharomyces:* a cytological study of a frozen-dried polyploid series. J. Bacteriol. 68, 514—529.
Nagano, T., 1959: Spermatogenesis of the domestic fowl studied with the electron microscope. Anat. Inst. Med. Fac. Okayama 164, 311—345. (In Japanese.)
— 1961: Personal communication.
— 1962: Observations on the fine structure of the developing spermatid in the domestic chicken. J. Cell Biol. 14, 193—205.
Nath, V., 1956: Cytology of spermatogenesis. Int. Rev. Cytol. 5, 395—453.
Nebel, B. R., and E. M. Coulon, 1962: The fine structure of chromosomes in pigeon spermatocytes. Chromosoma (Berlin) 13, 272—291.
Norman, W. W., 1896: Segmentation of the nucleus without segmentation of the protoplasm. Roux' Arch. Entwicklungsmechanik Org. 3, 106—126.
Östergren, G., A. Koopmans, and J. Reitalu, 1953: The occurence of the amphiastral type of mitosis in higher plants and the influence of aminopyrin on mitosis. Botaniska Notiser 4, 417—419.
— J. Molè-Bajer, and A. Bajer, 1960: An interpretation of transport phenomena at mitosis. Ann. N. Y. Acad. Sci. 90, 381—408.
Payne, F., 1927: Some cytoplasmic structures in the male germ cells of *Gelastocoris oculatus* (toad-bug). J. Morph. 43, 299—345.
Pease, D. C., 1941: Hydrostatic pressure effects upon the spindle figure and chromosome movement. I. Experiments on the first mitotic division of *Urechis* eggs. J. Morph. 69, 405—441.
— 1946: Hydrostatic pressure effects upon the spindle figure and chromosome movement. II. Experiments on the meiotic division of *Tradescantia* pollen mother cells. Biol. Bull. 91, 145—165.
Penrose, L. S., 1959: Self-reproducing machines. Sci. Amer. 200, 105—114.
Poglazov, B. F., 1961: The action of ATP and reducing agents on the achromatic apparatus of loach eggs (experiments in vivo). Tsitologiia 3, 204—206. (In Russian.)
Pollister, A. W., 1933: Notes on centrioles of amphibian tissue cells. Biol. Bull. 65, 529—545.
— and P. F. Pollister, 1943: The relation between centriole and centromere in atypical spermatogenesis of viviparid snails. Ann. N. Y. Acad. Sci. 45, 1—48.
Pontecorvo, G., 1958: Self-reproduction and all that. Symposia Soc. Exptl. Biol. 12, 1—5.
Porter, K. R., 1955: Changes in cell fine structure accompanying mitosis, in: Fine Structure of Cells. Sym. 8th Congress of Cell Biol. 236—250. New York: Interscience.
— and J. Blum, 1953: A study in microtomy for electron microscopy. Anat. Rec. 117, 685—712.
— and R. D. Machado, 1960: Studies on the endoplasmic reticulum. IV. Its form and distribution during mitosis in cells of onion root tip. J. Biophys. Biochem. Cytol. 7, 167—180.

RAPKINE, L., 1931: Sur les processus chemiques au cours de la division cellulaire. Ann. Physiol. Physicochem. Biol. 7, 382—418.

RAVEN, C. P., 1958: Morphogenesis: The analysis of molluscan development. New York: Pergamon Press, xii, 1—311.

RESENDE, F., 1947: Karyokinesis. Portug. Acta Biol. 2, 1—24.

RICHARDS, B. M., and A. BAJER, 1961: Mitosis in endosperm. Changes in nuclear and chromosome mass during mitosis. Exper. Cell Res. 22, 503—508.

RIS, H., 1955: Cell division, in: Analysis of Development (B. H. WILLIER, P. A. WEISS, and V. HAMBURGER, eds.), 1955. Philadelphia: W. B. Saunders Co. xii, 1—735.

ROTH, L. E., and E. W. DANIELS, 1962: Electron microscope studies of mitosis in amebae. II. The giant ameba Pelomyxa carolinensis. J. Cell Biol. 12, 57—78.

— S. W. OBETZ, and E. W. DANIELS, 1960: Electron microscopic studies on mitosis and amoeba. I. Amoeba proteus. J. Biophys. Biochem. Cytol. 8, 207—220.

RUSTAD, R. C., 1959 a: Further observations relating radiation-induced mitotic delay to centriole damage. Biol. Bull. 117, 437.

— 1959 b: The inhibition of mitosis in the sea urchin egg by acridine orange. Biol. Bull. 117, 437—438.

— 1959 c: Induction of multipolar spindles by single X-irradiated sperm. Exper. 15, 323.

— 1959 d: An interference microscopical and cytochemical analysis of local mass changes in the mitotic apparatus during mitosis. Exper. Cell Res. 16, 575—583.

— 1959 e: 96. Centriole damage: a possible explanation of radiation-induced mitotic delay. Radiation Res. 11.

— 1960: Changes in the sensitivity to ultraviolet-induced mitotic delay during the cell division cycle of the sea urchin egg. Exper. Cell Res. 21, 596—602.

— 1961 a: The centriole hypothesis of radiation-induced mitotic delay. Pathol. Biol. 9, 493—494.

— 1961 b: 130. The induction of mitotic delay by radiation and acridine orange. Radiation Res. 14.

RUTHMANN, A., 1958: The fine structure of the meiotic spindle of the crayfish. J. Biophys. Biochem. Cytol. 5, 177—180.

SAGER, R., 1965: On non-chromosomal heredity in microorganisms, in: Function and Structure in Microorganisms (M. R. POLLOCK and M. H. RICHMOND, eds.), London: Cambridge University Press, 324—342.

SAKAI, H., 1962 a: Studies on sulfhydryl groups during cell division of sea urchin egg. IV. Contractile properties of the thread model of KCl-soluble protein from the sea urchin egg. J. Gen. Physiol. 45, 411—425.

— 1962 b: Studies on sulfhydryl groups during cell division of sea urchin egg. V. Change in contractility of the thread model in relation to cell division. J. Gen. Physiol. 45, 427—438.

— and K. DAN, 1959: Studies on sulfhydryl groups during cell division of sea urchin egg. Exper. Cell Res. 16, 24—41.

SATIR, P., and B. SATIR, 1964: A model for ninefold symmetry in alpha-keratin and cilia. J. theor. Biol. 7, 123—128.

SATÔ, S., 1958: Electron microscope studies on the mitotic figure. I. Fine structure of the metaphase spindle. Cytol. (Tokyo) 23, 383—394.

— 1959: Electron microscope studies on the mitotic figure. II. Phragmoplast and cell plate. Cytol. (Tokyo) 24, 98—106.

— 1960: Electron microscopic studies on the mitotic figure. III. Process of spindle formation. Cytol. (Tokyo) 25, 119—131.

SAUAIA, H., and D. MAZIA, 1961: Action of colchicine on the mitotic apparatus. Pathol. Biol. 9, 473—476.

SCHRADER, F., 1941: The spermatogenesis of the earwig Anisolabis maritima Bon. with reference to the mechanism of chromosomal movement. J. Morph. 68, 123—147.

— 1953: Mitosis: The Movements of Chromosomes in Cell Division. 2nd ed. New York: Columbia University Press. xii, 1—170.

SCHREINER, A., und K. E. SCHREINER, 1905: Über die Entwicklung der männlichen Geschlechtszellen von Myxine glutinosa (L.) Arch. Biol. 21, 315—355.

SCHUEL, H., 1961: A study of the effects of urethane on the cleavage of the Chaetopterus egg. I. Inhibition of cleavage. Biol. Bull. 120, 384—400.

SCHULTZ-LARSEN, J., 1953: On the structure of the nuclear spindle. Acta Pathol. Microbiol. Scand. 32, 567—573.

Seaman, G. R., 1960: Large-scale isolation of kinetochores from the ciliated protozoan *Tetrahymena pyriformis*. Exper. Cell Res. 21, 292—302.
— 1962: Protein synthesis by kinetochores isolated from the protozoan *Tetrahymena*. Biochim. Biophys. Acta 55, 889—899.
Sears, E. R., and A. Camara, 1952: A transmissible dicentric chromosome. Genetics 37, 125—135.
Sentein, P., 1961 a: Action inhibitrice du phényluréthane sur la genèse des fibrilles pendant la seconde mitose de segmentation et accélération de la multiplication des centres cellulaires. Compt. rend. Soc. Biol. 155, 2418.
— 1961 b: L'action des antimitotiques pendant la segmentation de l'oeuf et le mécanisme de cette action. Pathol. Biol. 9, 445—466.
— 1962 a: Dissociation des différents mécanisms de la mitose de segmentation. Compt. rend. Acad. Sci., Paris, 254, 558—560.
— 1962 b: Le déterminisme des mitoses pluripolaires et leur mécanisme d'après l'action interrompue du phényluréthane sur l'oeuf d'urodèle. Chromosoma (Berlin) 13, 67—105.
Serra, J. A., and M. M. P. Seixas, 1962: On the existence of lipids in the centromere and the spindle. Rev. Portug. Zool. Biol. Geral. 3, 263—266.
Sharp, L. W., 1921: An Introduction to Cytology. New York: McGraw-Hill Book Co., Inc. xiii, 1—452.
— 1934: Introduction to Cytology. New York: McGraw-Hill Book Co., Inc. xiv, 1—567.
Shimamura, T., and T. Ōta, 1956: Cytochemical studies on the mitotic spindle and the phragmoplast of plant cells. Exper. Cell Res. 11, 346—361.
— — and T. Hishida, 1957: Cytochemical studies on the mitotic spindle. Symposia Soc. Cellular Chem. (Tokyo) 16, 24—41. (In Japanese with American summary.)
Sleigh, M. A., 1962: The Biology of Cilia and Flagella. Vol. 12 of Int. Series of Monographs on Pure and Applied Biology. Oxford: Pergamon Press, Ltd.
Sotelo, J. R., and O. Trujillo-Cenoz, 1958 a: Electron microscope study of the kinetic apparatus in animal sperm cells. Z. Zellforsch. 48, 565—601.
— — 1958 b: Electron microscope study on the development of ciliary components of the neural epithelium of the chick embryo. Z. Zellforsch. 49, 1—12.
Stafford, D. W., and R. M. Iverson, 1964: Radioautographic evidence for the incorporation of leucine-C^{14} into the mitotic apparatus. Science 143, 580—581.
— W. H. Sofer, and R. M. Iverson, 1964: Demonstration of polyribosomes after fertilization of the sea urchin egg. Proc. Nat. Acad. Sci. (Wash.) 52, 313—316.
Stanier, R. Y., and C. B. van Niel, 1962: The concept of a bacterium. Arch. Mikrobiol. 42, 17—35.
Stern, H., 1956: Sulfhydryl groups and cell division. Sci. 124, 1292—1293.
— 1958: Variations in sulfhydryl concentration during microsporocyte meiosis in the anthers of *Lilium* and *Trillium*. J. Biophys. Biochem. Cytol. 4, 157—161.
Stich, H., 1951: Das Vorkommen von Kohlenhydraten im Ruhkern und während der Mitose. Chromosoma (Berlin) 4, 429—438.
— 1954 a: Stoffe und Strömungen in der Spindel von *Cyclops strenuus*. Ein Beitrag zur Mechanik der Mitose. Chromosoma (Berlin) 6, 199—236.
— 1954 b: Der Einfluß von Giften auf die zur Meiose führenden Stoffwechselvorgänge bei *Sabellaria spinulosa*. Exper. 10, 184—185.
— and J. McIntyre, 1958: X-ray absorption studies on the nuclear protein and RNA content during the development of the mitotic apparatus. Exper. Cell Res. 14, 635—638.
Strasburger, E., 1897: Kerntheilung und Befruchtung bei *Fucus*. Jahrb. wiss. Bot. 30, 351—374.
Sturdivant, H. P., 1931: Central bodies in the sperm-forming divisions of *Ascaris*. Sci. 73, 417—418.
Swann, M. M., 1951 a: Protoplasmic structure and mitosis. I. The birefringence of the metaphase spindle and asters of the living sea-urchin egg. J. Exper. Biol. 28. 417—433.
— 1951 b: Protoplasmic structure and mitosis. II. The nature and cause of birefringence changes in the sea urchin at anaphase. J. Exper. Biol. 28, 434—444.
— 1954: The mechanism of cell division: experiments with ether on the sea urchin egg. Exper. Cell Res. 7, 505—517.
Swingle, W., 1897: Zur Kenntniss der Kern- und Zelltheilung bei den Sphacelariaceen. Jahrb. wiss. Bot. 30, 297—350.

SWINGLE, W. W., 1926: The germ cells of anurans. II. An embryological study of sex determination in *Rana catesbeiana*. J. Morph. 41, 441—546.

SZOLLOSI, D., 1964: The structure and function of centrioles and their satellites in the jellyfish *Phialidium gregarium*. J. Cell Biol. 21, 465—479.

TAYLOR, E., 1959: Dynamics of spindle formation and its inhibition by chemicals. J. Biophys. Biochem. Cytol. 6, 193—196.

— 1963: Relation of protein synthesis to the division cycle in mammalian culture cells. J. Cell Biol. 19, 1—18.

TSCHASSOWNIKOW, S., 1914: Über Becher- und Flimmerepithelzellen und ihre Beziehungen zueinander. Arch. mikr. Anat. 84, 150—174.

VAN BENEDEN, E., 1876: Contribution a l'historie de la vésicule germinative et du premier noyau embryonnaire. Bull. Acad. Roy. Belgique 41. (Cited in WILSON 1928.)

— and A. NEYT, 1887: Nouvelles recherches sur la fécondation et la division cellulaire chez l'*Ascaride mégalocéphale*. Bull. Ac. Roy. de Belgique (Brussels) 7. (Cited in WILSON 1928.)

WADA, B., 1950: The mechanism of mitosis based on studies of the submicroscopic structure and of the living state of the *Tradescantia* cell. Cytol. 16, 1—26.

WENT, H. A., 1959: Studies on the mitotic apparatus of the sea urchin by means of antigen-antibody reactions in agar. J. Biophys. Biochem. Cytol. 6, 447—455.

— 1960: Dynamic aspects of mitotic apparatus protein. Ann. N.Y. Acad. Sci. 90, 422—429.

— 1962: Structural modifications of the mitotic apparatus during the early cleavages in sand dollar eggs. Chromosoma (Berlin) 13, 219—242.

WILSON, E. B., 1897: Centrosome and middle-piece in the fertilization of the egg. Sci. N. S. 5, 390.

— 1898: The Cell in Development and Inheritance. 1st Ed. New York: The MacMillan Co. xvii, 1—377.

— 1901: A cytological study of artificial parthenogenesis in sea-urchin eggs. Arch. Entwicklungsmech. Org. (Berlin) 12. (Cited in WILSON 1928.)

— 1928: The Cell in Development and Heredity. 3rd Ed. New York: The MacMillan Co. xxxvii, 1—1232.

— 1930: The question of the central bodies. Sci. 71, 661—662.

— and A. F. HUETTNER, 1931: The central bodies again. Sci. 73, 447—448.

— and A. P. MATHEWS, 1895: Maturation, fertilization and polarity in the echinoderm egg. J. Morph. 10. (Cited in HERTWIG 1906, 270.)

WOLBACH, S. B., 1928: Centrioles and the histogenesis of the myofibril in tumors of striated-muscle origin. Anat. Rec. 37, 255—274.

WOODBURN, W. L., 1911: Spermatogenesis in certain Hepaticae. Ann. Bot. 25, 299—313.

— 1913: Spermatogenesis in *Blasia pusilla* L. Ann. Bot. 27, 93—101.

— 1915: Spermatogenesis in *Mnium affine* var. *ciliaris* (Grev.), C. M., Ann. Bot. 29, 441—456.

YATSU, N., 1904: Experiments on the development of egg fragments in *Cerebratulus*. Biol. Bull. 6, 123—136.

— 1905: The formation of centrosomes in enucleated egg-fragments. J. Exptl. Zool. 2, 287—312.

— 1908: Some experiments on cell-division in the egg of *Cerebratulus lacteus*. Ann. Zool. Japon. 6, 267.

ZIMMERMAN, A., 1960: Physico-chemical analysis of the isolated mitotic apparatus. Exper. Cell Res. 20, 529—547.

— and D. MARSLAND, 1964: Cell division: effects of pressure on the mitotic mechanisms of marine eggs (*Arbacia punctulata*). Exper. Cell Res. 35, 293—302.

ZIRKLE, R. E., R. B. URETZ, and R. H. HAYNES, 1960: Disappearance of spindles and phragmoplasts after microbeam irradiation of cytoplasm. Ann. N.Y. Acad. Sci. 90, 435—439.